Research Progress in
NANO AND INTELLIGENT
MATERIALS

Research Progress in
NANO AND INTELLIGENT MATERIALS

Edited By
A. K. Haghi, PhD

Professor, University of Guilan, Iran;
Editor-in-Chief, International Journal of
Chemoinformatics and Chemical Engineering

Apple Academic Press

TORONTO   NEW JERSEY

CRC Press
Taylor & Francis Group
6000 Broken Sound Parkway NW, Suite 300
Boca Raton, FL 33487-2742

Apple Academic Press, Inc
3333 Mistwell Crescent
Oakville, ON L6L 0A2
Canada

© 2012 by Apple Academic Press, Inc.
Exclusive worldwide distribution by CRC Press an imprint of Taylor & Francis Group, an Informa business

First issued in paperback 2021

No claim to original U.S. Government works

Version Date: 20120530

ISBN 13: 978-1-77463-194-2 (pbk)
ISBN 13: 978-1-926895-03-1 (hbk)

Visit the Taylor & Francis Web site at
http://www.taylorandfrancis.com

and the CRC Press Web site at
http://www.crcpress.com

For information about Apple Academic Press product
http://www.appleacademicpress.com

Contents

List of Contributors

Zahra Abadi
Department of Textile Engineering, University of Guilan, Rasht, Iran.

A. Afzali
Department of Textile Engineering, Faculty of Engineering, University of Guilan, Rasht, Iran.

Ali Benvidi
Department of Chemistry, Yazd University, Yazd, Iran.

Seyed M. Bidoki
Department of Textile Engineering, Yazd University, Yazd, Iran.

Mehdi Farokhi
National Cell Bank of Iran, Pasteur Institute of Iran, Tehran, Iran.
Department of Tissue Engineering and Cell Therapy, School of Advanced Medical Technologies, Tehran University of Medical Sciences, Tehran, Iran.

A. K. Haghi
Department of Textile Engineering, University of Guilan, Rasht, Iran.

A. A. Heidari
Department of Electrical and Computer Engineering, Yazd University, Yazd, Iran.

S. A. Hoseini
Textile Engineering Faculty, Isfahan University of Technology, Isfahan, Iran.

M. Kanafchian
University of Guilan, Iran.

Z. M. Mahdieh
University of Guilan, Rasht, Iran.

M. S. Motlagh
NanoRod Behan Co. (LTD), Guilan Science and Technology Park (GSTP), Rasht, Iran.

Fatemeh Mottaghitalab
National Cell Bank of Iran, Pasteur Institute of Iran, Tehran, Iran.
Department of Nanobiotechnology, Faculty of Basic Sciences, Tarbiat Modares University (TMU), Tehran, Iran.

V. Mottaghitalab
Department of Textile Engineering, Faculty of Engineering, University of Guilan, Rasht, Iran.

J. Nouri
Department of Textile Engineering, Yazd University, Yazd, Iran.

N. Piri
University of Guilan, Rasht, Iran.

M. Saberi
NanoRod Behan Co. (LTD), Guilan Science and Technology Park (GSTP), Rasht, Iran.

M. Sadrjahani
Textile Engineering Faculty, Isfahan University of Technology, Isfahan, Iran.

Mohammad Ali Shokrgozar
National Cell Bank of Iran, Pasteur Institute of Iran, Tehran, Iran.

List of Abbreviations

APS	Angular power spectrum
CDS	Clinical diagnostic station
CNTs	Carbon nanotubes
Co/PET	Cotton/Polyester
cv	Coefficient variation
DAQ	Data acquisition
dB	Decibels
DCM	Dichloromethane
DLS	Dynamic light scattering
DMEM	Dulbecco modified eagle's medium
DMF	Dimethylformamide
DSC	Differential scanning calirometry
ECG	Electrocardiogram
ECL	Electrochemiluminescent
ECM	Extracellular matrix
EDS	Energy dispersive spectrum
FBS	Fetal bovine serum
FTIR	Fourier transform infrared spectrometer
GPRS	General packet radio service
hp	Hewlett Packard
HP	High pass
HR	Heart rate
ICD	Implantable cardioverter defibrillator
IJP	Inkjet printing
LP	Low pass
MWNTs	Multi-walled nanotubes
NCBI	National Cell Bank of Iran
NR	Neutral red assay
PAN	Polyacrylonitrile
PBS	Phosphate buffered saline
PCB	Printed circuit board
PET	Polyester
PPSN	Poly-propylene spun-bond nonwoven
PSD	Pore size distribution
PVA	Polyvinyl alcohol

SCE	Saturated calomel electrode
SE	Shielding effectiveness
SEM	Scanning electron microscopy
SWNTs	Single-walled carbon nanotubes
TEM	Transmission electron microscope
TFA	Triflouroacetic acid
TPS	Tissue culture polystyrene
WLAN	Wireless local area networks
WSN	Wireless sensor networks

Preface

The collection of topics in this book aims to reflect the diversity of recent advances in Nano and Intelligent Materials with a broad perspective which may be useful for scientists as well as for graduate students and engineers. This new book presents leading-edge research from around the world in this dynamic field.

The book offers scope for academics, researchers, and engineering professionals to present their research and development works that have potential for applications in several disciplines of Nano and Intelligent Materials. Contributions ranged from new methods to novel applications of existing methods to gain understanding of the material and/or structural behavior of new and advanced systems.

Contributions are sought from many areas of Nano and Intelligent Materials in which advanced methods are used to formulate (model) and/or analyze the problem. In view of the different background of the expected audience, readers are requested to focus on the main ideas, and to highlight as much as possible the specific advantages that arise from applying modern ideas. A chapter may therefore be motivated by the specific problem, but just as well by the advanced method used which may be more generally applicable.

Nanotechnology comprises the technological development in the nanometric scale. An interdisciplinary area, it is included in the field of nanoscience, focused on advances involving chemistry, physics, biology and computer science. Governed by the laws of quantum physics, the universe of nanoparticles makes materials behave very differently from usual, on a normal scale. For example, materials that are naturally not conductors my become semiconductors on a nanometric scale.

Known as intelligent materials, in nanotechnological treatment they present entirely new properties, or combined with the originals they have multiple or adaptable functions and allow interfacing with organic and inorganic components, amongst other characteristics.

From the nanoscale to the macroscopic scale, intelligent materials are triggering a response across both dimensions and scientific disciplines. World class, leading experts in the fields of chemistry, physics and engineering have contributed to Intelligent Materials, highlighting the importance of smart material science in the 21st century. In this exceptional text the expertise of specialists across the globe is drawn upon to present a truly interdisciplinary outline of the topic. Covering both a bottom-up chemical, and top-down engineering approach to the design of intelligent materials the contributors of the articles are bridging a vital gap between various scientific authorities. The influence of current research in this field on future technology is undisputed and potential applications of intelligent materials span nanoscience, nano technology, medicine, engineering, biotechnology, pharmaceutical and many other industries. This is an authoritative introduction to the most recent developments in the area, which will provide the reader with a better understanding of the almost unlimited opportunities in the progress and design of new intelligent materials.

Nano-carbon materials: Nano-carbon materials include carbon nanotubes, vapor grown carbon fibers, also known as nano-carbon fiber, diamond-like carbon. Carbon nanotubes have unique pore-like Structure , the use of the structural characteristics of the drug stored in carbon nanotubes and through a certain mechanism to stimulate the release of drugs to control drug into reality. In addition, the carbon nanotubes can be used to composite enhancer, electron probe (such as observations of the AFM probe protein structure, etc.) or display and field emission tip. Carbon fiber is usually transition metals Fe, Co, Ni and its alloys as the catalyst, to a low-carbon hydrocarbon compounds as carbon source, hydrogen as the carrier, in the 873 K ~ 1473 K temperatures generated, with extraordinary properties and good biological compatibility, in the medical field, have broad application prospects. Diamond-like carbon (referred to as DLC) is a diamond structure with a large number of C-C bond of the hydrocarbon polymers, can be a plasma or an ion beam deposited on surfaces to form nano-structured film, with excellent bio-compatibility, In particular, the blood compatibility. Sources, however, when compared with other materials, diamond-like carbon surfaces on the lower level of fibrinogen adsorption on the adsorption of albumin increased, blood vessels to reduce intimal hyperplasia, which diamond-like carbon films have an important in cardiovascular clinical application value.

Nano-polymer materials: Nano-polymer materials, also known as super-polymer nano-particles or polymer particles, the particle size scale in the range of 1 nm ~ 1000 nm. Such particles have a gel, stability and excellent absorption properties can be used for drugs, gene delivery and drug release carrier, as well as immunoassay, interventional treatment and so on.

Nanocomposites: At present, the Research and Development of inorganic-inorganic, organic-inorganic, organic-organic and biological activity-non-biological activity of nano-structured composite materials is to obtain a new generation of high performance functional composite materials, new ways, and gradually to the intelligent direction, In the light, heat, magnetism, force, sound and so on with singular characteristics, and thus in tissue repair and transplantation, and many other areas have broad application prospects. Abroad, have been prepared ZrO_2 toughened alumina nano-composite materials, artificial materials with this hip and knee implants the life of up to 30 years. Studies have shown that nano-hydroxyapatite collagen material is also a good bone tissue Engineering scaffold material. In addition, nano-hydroxyapatite nano-particles made of anti-cancer drugs, can kill cancer cells, effectively inhibit tumor growth, while the normal tissue intact, the results of this study attracted international attention.

Intelligent materials: An active structure consists of a structure provided with a set of actuators and sensors coupled by a controller; if the bandwidth of the controller includes some vibration modes of the structure, its dynamic response must be considered. If the set of actuators and sensors are located at discrete points of the structure, they can be treated separately. The distinctive feature of *smart* structures is that the actuators and sensors are often distributed and have a high degree of integration inside the structure, which makes a separate modelling impossible.

Moreover, in some applications like vibroacoustics, the behavior of the structure itself is highly coupled with the surrounding medium; this also requires a coupled modelling.

From a mechanical point of view, classical structural materials are entirely described by their elastic constants relating stress and strain, and their thermal expansion coefficient relating the strain to the temperature. *Smart materials* are materials where strain can also be generated by different mechanisms involving temperature, electric field or magnetic field, etc. as a result of some coupling in their constitutive equations. The most celebrated smart materials are briefly described below:

Shape Memory Alloys (SMA): SMAs allow one to recover up to 5% strain from the phase change induced by temperature. Although two-way applications are possible after education, SMAs are best suited for one-way tasks such as deployment. In any case, they can be used only at low frequency and for low precision applications, mainly because of the difficulty of cooling. Fatigue under thermal cycling is also a problem. SMAs are little used in vibration control.

Piezoelectric materials: They have a recoverable strain of 0.1% under electric field; they can be used as actuators as well as sensors. They are two broad classes of piezoelectric materials used in vibration control: ceramics and polymers. The piezoplymers are used mostly as sensors, because they require high voltages and they have a limited control authority; the best known is the polyvinylidene fluoride (PVF$_2$). Piezoceramics are used extensively as actuators and sensors, for a wide range of frequency including ultrasonic applications; they are well suited for high precision i the nanometer range (1 nm = 10^{-9} m). The best known piezoceramic is the Lead Zirconate Titanate (PZT).

Magnetostrictive materials: Magnetostrictive materials have a recoverable strain of 0.15% under magnetic field; the maximum response is obtained when the material is subjected to compressive loads. Magnetostricitive actuators can be used as load carrying elements (in compression alone) and they have a long life time. They can also be used in high precision applications.

The range of available devices to measure position, velocity, acceleration and strain is extremely wide, and there are more to come, particulary in optomechanics. Displacements can be measured with inductive, capacitive and optical means (laser interferometer); the latter two have a resolution in the nanometer range. Piezoelectric accelerometers are very popular but they cannot measure a d.c. component. Strain can be measured with strain gages, piezoceramics, piezopolymers and fiber optics. The latter can be embedded in a structure and give a global average measure of the deformation.

Smart materials and structures is an emerging technology with numerous potential applications in industries as diverse as consumer, sporting, medical and dental, computer, telecommunications, manufacturing, automotive, aerospace, as well as civil and structural engineering. Smart materials, similar to living beings, have the ability to perform both sensing and actuating functions and are capable of adapting to changes in the environment. In other words, smart materials can change themselves in response to an outside stimulus or respond to the stimulus by producing a signal of some sort. Hence, smart materials can be used as "sensors", "actuators" or in some cases as "self-sensing actuators" or "senoricactuators" in general. By utilizing these materials, a complicated part in a system consisting of individual structural, sensing and actuating components can now exist in a single component, thereby reducing overall size and complexity of the system. However, smart materials will never replace systems fully; they usually are part of some smart systems. Examples include smart airplane wings achieving greater fuel efficiency by altering their shape in response to air pressure and flying speed, vibration-damping systems for large civil engineering structures and automobile suspension systems, smart tennis rackets having rapid internal adjustments for overhead smashes and delicate drop shots, smart water purification systems for sensing and removing noxious pollutants, as well as smart medical systems treating diabetes with blood sugar sensors and insulin delivery pumps, to name a few.

Smart structures refer to those structures incorporating smart materials for sensors and actuators, structural identification techniques to obtain analytical/numerical models, and intelligent electronics and control methodologies to improve the structronic response. These structures have the ability to beneficially respond to internal and external simulations through the sensing, controlling and actuating of their structronic response. The design and implementation of smart structures necessitate the integration of 1) smart sensors and actuators, 2) structural analysis and design, and 3) intelligent electronics and control techniques. The field of smart structures is one of several areas of science (i.e., physics) and engineering where knowledge is distributed in several disciplines and requires integration into a single functional unit.

This book also deals with integrate research results with curriculum development for the benefit of students in physics, materials science and engineering, civil and structural engineering, mechanical and aerospace engineering, industrial and systems engineering, as well as electrical and electronic engineering.

The book also provides students with:

1. the fundamentals of smart materials, devices and electronics, in particular those related to the development of smart structures and products; and
2. the skills, knowledge and motivation in the design, analysis and manufacturing of smart structures and products.

On completion of this subject, students should be able to:

1. understand the physical principles underlying the behavior of smart materials;
2. understand the engineering principles in smart sensor, actuator and transducer technologies;
3. use principles of measurement, signal processing, drive and control techniques necessary to developing smart structures and products; and
4. appreciate and suggest improvement on the design, analysis, manufacturing and application issues involved in integrating smart materials and devices with signal processing and control capabilities to engineering smart structures and products.

I would like to express my deep appreciation to all the authors for their outstanding contribution to this book and to express my sincere gratitude for their generosity. All the authors eagerly shared their experiences and expertise in this new book. Special thanks go to the referees for their valuable work.

— **A. K. Haghi**

Chapter 1

Updates on PAN Monofilament in Nanoscale

M. Sadrjahani, S. A. Hoseini, and V. Mottaghitalab

INTRODUCTION

With potential applications ranging from protective clothing and filtration technology to reinforcement of composite materials and tissue engineering, nanofibers offer remarkable opportunity in the development of multifunctional material systems. The emergence of various applications for nanofibers is stimulated from their outstanding properties such as very small diameters, huge surface area per mass ratio and high porosity along with small pore size. Moreover, the high degree of orientation and flexibility beside superior mechanical properties are extremely important for diverse applications [1–3]. In this study, aligned and molecularly oriented polyacrylonitrile (PAN) nanofibers were prepared using a novel technique comprise two needle with opposite voltage and a rotating drum for applying take-up mechanism. The electrospinning process was optimized for increasing of productivity and improving the mechanical properties through controlling internal structure of the generated fibers.

BACKGROUND

Electrospinning is a sophisticated technique that relies on electrostatic forces to produce fibers in the nano to micron range from polymer solutions or melts. In a typical process, an electrical potential is applied between droplet of a polymer solution, or melt, held at the end of a capillary tube and a grounded target. When the applied electric field overcomes the surface tension of the droplet, a charged jet of polymer solution is ejected. The trajectory of the charged jet is controlled by the electric field.

The jet exhibits bending instabilities due to repulsive forces between the charges carried with the jet. The jet extends through spiraling loops, as the loops increase in diameter the jet grows longer and thinner until it solidifies or collects on the target [3]. The fiber morphology is controlled by the experimental design and is dependent upon solution conductivity, solution concentration, polymer molecular weight, viscosity, applied voltage, and distance between needle and collector [2, 3] due to initial instability of the jet, fibers are often collected as randomly oriented structures in the form of nonwoven mats, where the stationary target is used as a collector. These nanofibers are acceptable only for some applications such as filters, wound dressings, tissue scaffolds, and drug delivery [4]. Aligned nanofibers are another form of collected nanofibers that can be obtained by using rotating collector or parallel plates [2, 3]. Recent studies have shown that aligned nanofibers have better molecular orientation and as a result improved mechanical properties than randomly oriented nanofibers [3, 5, 6]. These nanofibers can be used in applications such as composite reinforcement and device manufacture [4]. Moreover, the aligned nanofibers are better suited for thermal or drawing treatment (the methods of collecting aligned nanofibers can be utilized for preparing of carbon nanofibers from electrospun PAN nanofibers precursor) [7]. Recently, PAN nanofibers have attracted a lot of interest as a precursor of carbon nano-fibers. Fennessey and his coworkers prepared tows of unidirectional and molecularly oriented PAN nanofibers using a high speed, rotating take up wheel. A maximum chain orientation parameter of 0.23 was determined for fibers collected between 8.1 and 9.8 m/s. The aligned tows were twisted into yarns, and the mechanical properties of the yarns were determined as a function of twist angle. Their yarn with twist angle of 11° had initial modulus and ultimate strength of about 5.8 GPa and 163 MPa, respectively [3]. Zussman et al. have demonstrated the use of a wheel-like bobbin as the collector to position and align individual PAN nanofibers into parallel arrays. They are obtained Herman's orientation factor of 0.34 for nanofibers collected at 5 m/s. [5]. Gu et al. collected aligned PAN nanofibers across the gap between the two grounded stripes of aluminum foil. They reported the increase of orientation factor from 0 to 0.127 and expressed to improve mechanical properties in particular the modulus of the resultant carbon fibers [6].

EXPERIMENTAL

Materials
Industrial PAN was provided by Iran Polyacryle and dimethylformamide (DMF) was purchased from Merck. The weight average molecular weight ($W\,M$) and the number average molecular weight ($n\,M$) of PAN were $W\,M$ =100,000 g/mol and $n\,M$ =70,000 g/mol. All solutions of PAN in DMF were prepared under constant mixing by mag-netic mixer at room temperature.

Electrospinning Setup
The electrospinning apparatus consist of a high voltage power supply, two syringe pumps, two stainless steel needles (0.7 mm OD) and a rotating collector with variable surface speed which is controlled by an Inverter (Fig. 1.1). In this setup unlike the

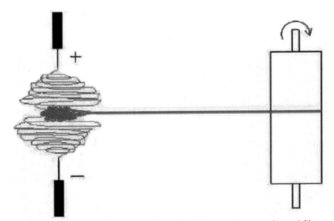

Figure 1.1. Schematic electrospinning setup for collection continuous aligned fibers.

conventional technique, two needles were installed in opposite direction and polymer solutions were pumped to needles by two syringe infusion pumps with same feed rate. The flow rate of solutions to the needle tip is maintained so that a pendant drop remains during electrospinning. The horizontal distance between the needles and the collector was several centimeters. When high voltages were applied to the needles with opposite voltage, jets were ejected simultaneously. Then the jets with opposite charges attracted each other, stuck together and a cluster of fibers formed. For collecting aligned nanofibers, the cluster of fibers formed between the two needles was towed manually to the drum is rotated from 0 to 3000 rpm.

Microscopy

Electrospun nanofibers were observed by scanning electron microscopy (SEM) and optical microscopy. Samples were mounted onto SEM plates; sputter coated with gold, and examined using a Philips XL-30 electron microscope to determine fiber diameters. A Motic optical microscope was used to capture images for alignment analysis. Fibers alignment was analyzed using image processing technique by Fourier power spectrum method as a function of take up speed. All images used in the process were obtained using the Motic optical microscopy at a resolution of 640 × 480 pixels in 1000 × magnifications. The number of captured images was 30 at each of take up speeds.

FTIR

Dried, electrospun fiber bundles were examined using a Bomen MB-Series 100 infrared spectrometer (FTIR) to measure crystallization index of PAN nanofibers as a function of collection speed. FTIR spectra were recorded over the range of 400–4000 cm^{-1} with 21 scans/min. The crystallization index (A1730/A2240) for PAN fibers was acquired by rationing the absorbance peak areas of nitrile (2240 cm^{-1}) and carbonyl (1730 cm^{-1}) groups. The relation between degree of crystallinity and A1730/A2240 is indirect (i.e., the larger A1730/A2240, the less crystalline fibers are) [8].

Raman Spectroscopy

Raman spectroscopy was used for the survey of molecular orientation variability with take up speed. Raman spectra were obtained with the Thermo Nicolt Raman spectrometer model Almega Dispersive 5555. The spectra were collected with a spectral resolution of 2-4260 cm^{-1} in the backscattering mode, using the 532 nm line of a Helium/Neon laser. The nominal power of the laser was 30 mW. A Gaussian/Lorentzian fitting function was used to obtain band position and intensity. The incident laser beam was focused onto the specimen surface through a 100 × objective lens, forming a laser spot of approximately 1 μm in diameter, using a capture time of 50s. The analysis of PAN nanofiber orientation using polarized Raman spectroscopy was carried out based on a coordination system defined in Fig. 1.2. The nanofiber axis is defined as Z and molecular chain are oriented at angle $\theta°$ with respect to the Z axis. The nanofiber is mounted on the stage of the Raman microscope such that the incident laser comes in along the x' axis. The angle between the polarization plane and the nanofiber axis is ψ. The orientation studies were performed when fibers were at ψ = 0° and ψ = 90° to the plane of polarization of the incident laser.

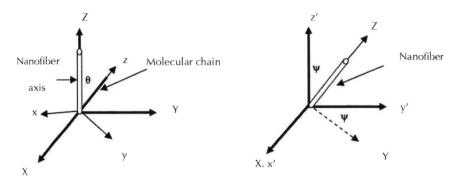

Figure 1.2. Demonstration of molecular chain coordination (xyz) in nanofiber (XYZ coordinates) and the nanofiber arrangement in reference frame of the Raman sample stage (x'y'z' coordinates).

Mechanical Testing

The mechanical behavior of bundles of aligned PAN nanofibers was examined using the Zwick 1446-60 with a crosshead speed of 60 mm/min and gauge length of 25 mm under standard conditions. All samples were included in standard container during 24 h under temperature of 25°C and relative humidity of 65% before tested. The initial modulus, stress, and strain of samples were determined.

Differential Scanning Calirometry (DSC)

DSC curves of electrospun fibers were obtained using a DSC 2010V4/4E by heating from 30 to 300°C at a heating rate of 10°C/min. The effect of take up speed on the Tg temperature and evolved heat per time was examined.

RESULTS AND DISCUSSION

Productivity

In this study, fibers were electrospun in the aligned form by using a simple and novel method which manufacture well-aligned polymer nanofibers with infinite length and over large collector area [4]. In this technique, the electric field only exists between two syringe needles and thus, the movement of nanofibers in drawing area (distance between the needles and the collector) is due to mechanical force caused by rotating drum and electric field do not has any portion at moving nanofibers from needles to collector. This is likely caused rupture of fibers in this technique. The concentration, the applied voltage, distance between two needles and distance between the syringe needles and the rotating drum were adjusted in order to obtain optimum and productive conditions. The conditions pertinent to minimum number of rupture for nanofibers prepared at different concentrations are shown in Table 1.1.

Table 1.1. The conditions obtained to generate nanofibers with minimum amount of rapture.

Solution concentration (Wt%)	Applied voltage (Kv)	Feed rate (mL/h)	Distance between two needles (cm)	Distance between needles and collector (cm)	Total number of rupture (at 15 minutes)
13	10.5	0.293	13	20	12
14	11	0.293	13	20	6
15	11	0.293	15	20	6

PAN nanofibers prepared at 14 and 15 wt% concentrations have lower rupture which can be due to high chain entanglement in these concentrations. Figure 1.3 shows SEM images of PAN nanofibers electrospun in obtained optimum conditions. The generated nanofibers have uniform structure and beaded fiber do not observes. As expected, the average diameter of nanofibers increased by increasing solution concentration (Table 1.2).

Table 1.2. Average diameter of electrospun nanofibers at different solution concentrations.

Concentration \ Diameter (nm)	Average	Coefficient Variation (CV%)	$\bar{x} \pm sd$ (nm)
13 wt%	323.45	9.59	323.45 ± 31.03
14 wt%	394.19	7.32	394.19 ± 28.84
15 wt%	404.67	10.82	404.67 ± 43.81

This is probably due to the greater resistance of the solution to be stretched by the charges on the jet with increasing the volume percent of solid in the solution and viscosity [2]. The prepared nanofibers at 14 wt% concentration have better uniform structure since the percent of coefficient variation (cv%) of diameter at 14 wt% is less than other concentrations. The viscosity is high at 15 wt% concentration and result in difficultly pumping the solution through syringe needle and solution dry at the tip of needle somewhat. By considering this thought and also the value of rupture, uniformity of diameter, and suitable diameter in nanoscale range, the conditions acquired at the 14 wt% concentration was chosen as a desired option for producing PAN nanofibers.

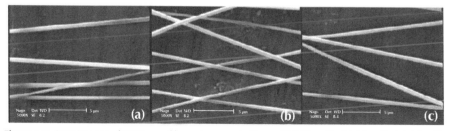

Figure 1.3. SEM images of PAN nanofibers at concentrations of (a) 13 wt%; (b) 14 wt%; (c) 15 wt%.

Alignment

Analysis of fibers alignment was carried out by obtaining angular power spectrum (APS) of nanofibers collected at different take up speeds from 22.5 m/min to 67.7 m/min. The plot of normalized APS (ratio of intensity of the APS to the corresponding mean intensity of the Fourier power spectrum) versus angle was used for calculating degree of alignment (Fig 1.4). In this plot area of peak at angle of 90° shows density of aligned nanofibers in the rotation direction of the collector. For relative comparing among samples, the ratio of Apeak/ATotal (area of peak at angle of 90°/total area of APS plot) was utilized. The alignment of the collected fibers is induced by the rotation of the target and improves as the surface velocity of the target is increased (Table 1.3). As the rotation speed increases, the effective draw (difference between surface velocity of drum and final velocity of fiber) is increased resulting in better alignment of the collected fiber and less deviation between the fiber and rotation direction. Also, the results show that more increase of take up speed causes no further increase of alignment. The maximum amount of degree of alignment obtained 37.5% at take up speed of 59.5 m/min.

Table 1.3. The degree of alignment of the collected fibers.

Take-up (m/min)	22.5	31.6	40.6	49.6	59.5	67.7
Degree of alignment (%)	24.59 ± 3.97	34.4 ± 5.29	32.72 ± 7.65	29.48 ± 5.97	37.53 ± 5.26	29.43 ± 7.04

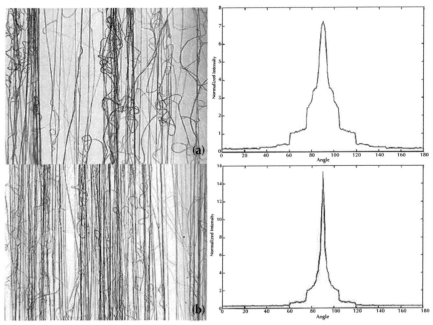

Figure 1.4. Optical micrograph of electrospun PAN nanofibers with corresponding normalized APS at take-up speeds of (a) 22.5 m/min; (b) 59.5 m/min.

Crystallization Index

The crystallization index (A1730/A2240) was calculated from FTIR spectra of PAN nanofibers collected at different take-up speeds from 22.5 m/min to 67.7 m/min in optimum conditions. The obtained results (Fig. 1.5) and performance ANOVA statistic analysis over them show that the increase of surface velocity has no effect on the crystallization index and as a result the crystallinity of PAN nanofibers has not varied.

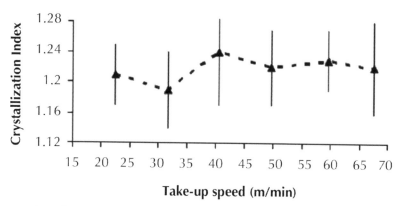

Figure 1.5. Crystallization Index of PAN nanofibers versus take-up speed.

Molecular Orientation

The molecular orientation of the nanofibers was examined by Raman Spectroscopy. Raman spectra were collected from bundles of fibers electrospun at 11 kV from 14 wt% PAN in DMF solutions collected onto a drum rotating with a surface velocity between 22.5 m/min and 67.7 m/min. The main difference among different molecular structures of PAN fibers usually arise in the region of 500–1500 cm^{-1} which is called as finger point region [9]. In this region, the peaks over the ranges of 950–1090 cm^{-1} and 1100–1480 cm^{-1} are common [9,10], be observed at Raman spectra of generated PAN nanofibers (Fig. 1.6).

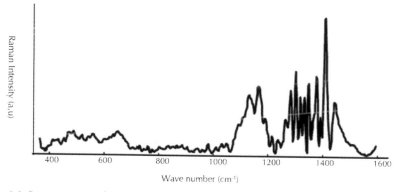

Figure 1.6. Raman spectra of PAN nanofibers.

Figure 1.7 shows the Raman spectra of different samples of the nanofibers using VV configuration for different amounts of ψ. Raman spectra were obtained in two directions, parallel (ψ = 90°) and vertical direction (ψ = 90°) with respect to the polarization plane.

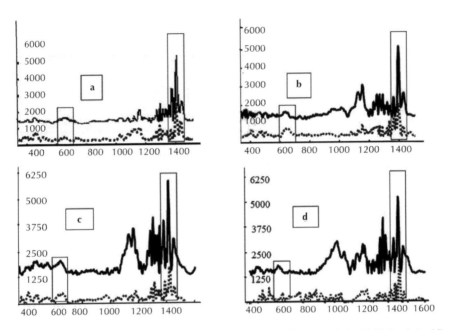

Figure 1.7. The Raman spectra under VV mode, (a) 22.5 m/min, (b) 49.6 m/min, (c) 59.5 m/min, (d) 67.7 m/min. From top to bottom, the angle between fiber axis and polarization plane is 0o and 90o.

Compared to Peak enhanced in 600 cm⁻¹ with constant intensity in different polarization angle, the intensity at 1394 cm⁻¹ monotonically decreases with increasing of ψ. The intensity dependence of peak enhanced in 1394 cm⁻¹ to the angle of fiber and polarization plane can be considered as powerful tool for determination of nanofibers orientation. Other peaks that enhanced between 1594 cm⁻¹ and 1169 cm⁻¹ did not decreased significantly, except the peak enhanced at 1190 cm⁻¹ and 1454 cm⁻¹. It is worth noting that trend in peak intensity observed for different sample shown in Fig. 1.8 in VH configuration can also be attributed to different orientation magnitude. When the nanofibers were examined using the VH configuration, the intensity dependence of enhanced peak at 1394 cm⁻¹ on ψ showed a different trend for different sample that cannot be directly correlated to degree of orientation. The higher intensity obtained for sample d, however the intensity value decreases for samples c, b, and a, respectively as shown in Fig. 1.8.

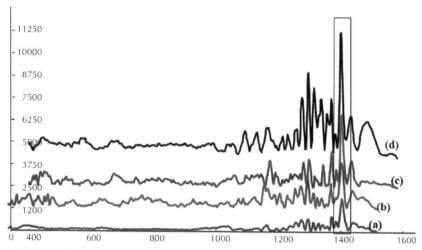

Figure 1.8. The Raman spectra under VH configuration at zero angle between fiber axis and polarization plane, (a) 22.5 m/min, (b) 49.6 m/min, (c) 59.5 m/min, and (d) 67.7 m/min.

According to intensity ratios shown in Table 1.4, Raman spectra show a much stronger orientation effect in the sample c compared with the other spun nanofibers. It is clear that the higher drum take up speed is mostly responsible for the orientation of molecular chain in the fiber direction. Therefore, as expected, the low orientation dependence of Raman modes was observed in lower take up speed.

Table 1.4. The intensity ratios derived from Fig. 1.7 and Fig. 1.8 based on shown values of enhanced peak at 1394 cm^{-1}.

Sample	I_{VV0}	I_{VV90}	I_{VH0}	I_{VV90}/I_{VH0}	I_{VV0}/I_{VH0}	I_{VV0}/I_{VV90}
A	4400	1900	1600	1.1875	2.75	2.31
B	3600	1500	2400	0.625	1.5	2.40
C	4000	1500	4277	0.35071	0.93523	2.67
D	3800	2200	6350	0.34646	0.59843	1.73

The mathematic formulation of Raman intensity in VV and VH modes are represented by the following expressions [11, 12]:

$$I^{vv}(\psi)\alpha\left(\cos^4\psi - \frac{6}{7}\cos^2\psi + \frac{3}{35}\right)\langle P_4(\cos\theta)\rangle$$
$$+\left(\frac{6}{7}\cos^2\psi - \frac{2}{7}\right)\langle P_2(\cos\theta)\rangle + \frac{1}{5} \qquad \text{Equation 3.1}$$
$$I^{VH}(\psi)\alpha\left(-\cos^4\psi - \frac{4}{35}\right)\langle P_4(\cos\theta)\rangle$$
$$+\frac{1}{21}\langle P_2(\cos\theta)\rangle + \frac{1}{15} \qquad \text{Equation 3.2}$$

The orientation order parameters of <P2 (cos θ)> and <P4 (cos θ)> which are, respectively, the average values of P2 (cos θ) and P4 (cos θ) for the SWNTs bulk product. The Pi(cos θ) is the Legendre polynomial of degree i which is defined as P2(cos θ)= (3 cos 2 θ-1)/2 and P4(cos θ)= (35 cos 4 θ – 30 cos 2 θ + 3)/8 for the second and fourth degree, respectively. More specifically the <P2(cos θ)> is known as the Herman's orientation factor (f) which varies between values of 1 and 0 corresponding, respectively, to nanotubes fully oriented in the fiber direction and randomly distributed [11].

$$f = \frac{3\langle \cos^2 \theta \rangle - 1}{2} \qquad\qquad Equation\ 3.3$$

The orientation factors can be determined by solving of following simultaneous algebraic equations given in 3.4 and 3.5. These equations are obtained from equations 3.1 and 3.2 by dividing of both side of these equations and substitution of angles of 0 and 90 degrees for ψ.

$$\frac{I_{G,RBM}^{VV}(\psi = 0)}{I_{G,RBM}^{VH}(\psi = 0)} = -\frac{24\langle P_4(\cos\theta)\rangle + 60\langle P_2(\cos\theta)\rangle + 21}{12\langle P_4(\cos\theta)\rangle - 5\langle P_2(\cos\theta)\rangle - 7} \qquad Equation\ 3.4$$

$$\frac{I_{G,RBM}^{VV}(\psi = 90)}{I_{G,RBM}^{VH}(\psi = 0)} = \frac{-9\langle P_4(\cos\theta)\rangle + 30\langle P_2(\cos\theta)\rangle + 21}{12\langle P_4(\cos\theta)\rangle - 5\langle P_2(\cos\theta)\rangle - 7} \qquad Equation\ 3.5$$

The left hand side terms of equations of 3.4 and 3.5 are the depolarization ratios that can be experimentally determined. As it can be seen only the intensity at 0ᵗ and 90ᵗ are required to determine the <P2 (cos θ)> and <P4 (cos θ)> for a uniaxially oriented nanofibers. Results calculated from equation 3.4 and 3.5 for Herman orientation factor for different sample shows a range between 0.20 and 0.25 at take up speeds of 67.7 m/min and 59.5 m/min, respectively (Fig. 1.9).

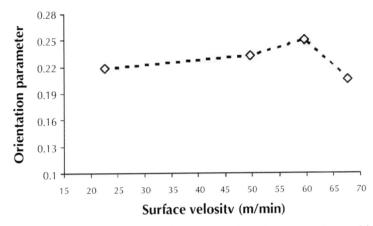

Figure 1.9. Orientation parameter versus take-up speed of rotating drum determined by Raman Spectroscopy.

As results show, the maximum chain orientation parameter yield at speed of 59.5 m/min and further increase of it causes loss of molecular orientation which corresponding to other studies [3]. As regards a distance between the needles and the collector is fixed, it may be due to decrease of drawing time of nanofibers with increasing take up speed and as a result low sufficient opportunity for arranging molecular chains in draw direction. Therefore, it appears that applying high draw ratio at short time has no significant effect on molecular orientation. Comparing this results with orientation factor of 0.66 and 0.52, which have been observed for commercial wet-spun acrylic fibers and melt-spun acrylic fibers, respectively [13], it can be stated that electrospun PAN nanofibers have lower molecular orientation than commercial fibers.

In general, based on the data of orientation factor for different sample, it can be clearly claimed that the increase of surface velocity of collector has a quite positive impact on molecular orientation of PAN nanofibers.

Mechanical Properties

Unidirectional bundles of aligned PAN nanofibers prepared from 14 wt% PAN in DMF solutions electrospun at 11 kV onto a drum rotating with a surface velocity ranging from 22.5 m/min to 67.7 m/min. The average linear density of the bundles was 176 dens. The stress-strain behavior of the bundles was examined and the modulus, ultimate strength, and elongation at the ultimate strength were measured as a function of take up speed. The initial modulus and ultimate strength increase gradually with take up speed from 1.3 GPa (12.31 g/den) and 61.7 MPa (0.577 g/den) at a surface velocity of 22.5 m/min to 4.2 GPa (39.43 g/den) and 73.7 MPa (0.694 g/den) with a liner velocity of 59.5 m/min, respectively.

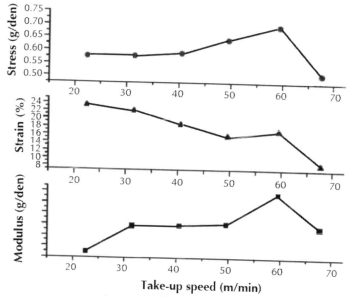

Figure 1.10. Stress, strain and modulus of PAN nanofibers versus take-up speed.

The modulus and ultimate strength of the bundles decreased with take up speed greater than 59.5 m/min (Fig. 1.10). The elongation at ultimate strength followed an inverse trend; it decreased from 23.3% to 8.4% with increasing take up speed from 22.5 m/min to 67.7 m/min, respectively.

Analysis of Mechanical Properties of PAN Nanofiber Bundles

In investigation of the results obtained for mechanical properties of PAN nanofibers bundles, the effect of parameters such as internal structure and arrangement of fibers with together (spatial orientation) were analyzed. As was shown, the increase of take up speed did not affect on crystallization of PAN nanofibers, therefore, this parameter can not be had main role in variation trend of fiber bundles strength. Figure 1.11-a, b show the plot of stress versus degree of alignment and molecular orientation parameter, respectively. Degree of alignment determines the number of fibers subjected in tension direction and increases stress of bundles of fiber. The positive correlation factor of 0.53 (acquired by SAS statistic software) between stress and alignment demonstrates direct relation of them, but the low amount of this coefficient offers low correlation. On the other hand, the positive and high correlation factor of 0.99 concludes liner and good correlation between stress and molecular orientation parameter. Thus, it can be stated that the molecular orientation performs main and important task in response of mechanical properties and particularly stress of PAN nanofibers.

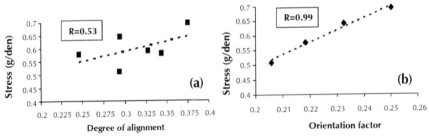

Figure 1.11. The stress of PAN nanofibers versus (a) degree of alignment; (b) molecular orientation parameter.

Thermal Properties of PAN Nanofibers

The obtained results from DSC curves of PAN nanofibers (Fig. 1.12) collected onto a rotating drum with surface velocity between 22.5 m/min and 67.7 m/min are summarized in Table 1.5.

The glass transition temperature (Tg) of PAN nanofibers has no considerable change with increasing take up speed and approximately is equal with 100°C. A high exothermic peak existed in DSC curve of PAN nanofibers is in relation to occurrence of dehydrogenation and cyclization reactions [14, 15, 16]. These chemical reactions must be conducted so to promote a slow release of heat, because, a rapid release of heat can cause a loss of molecular orientation and melting of the fiber, which will have a detrimental effect on the mechanical properties of the final fiber [14, 15]. The minimum rates of released heat yielded for PAN nanofibers collected at liner speed of 67.7

m/min and 59.5 m/min. Therefore, the use of above take up speeds can be suitable for thermal treatment of PAN nanofibers.

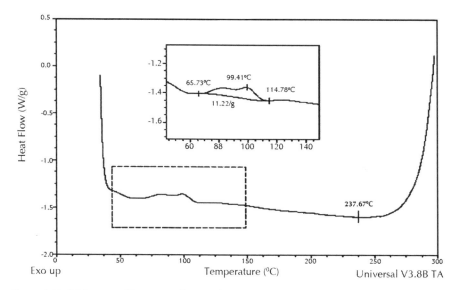

Figure 1.12. DSC curve of PAN nanofiber at take-up speed of 59.5 m/min

Table 1.5. Thermal properties of PAN nanofibers collected at different take-up speeds.

Surface velocity (m/min)	22.5	31.6	40.6	49.6	59.5	67.7
T_g(°C)	100.9	101.5	102	100.6	99.4	100.5
$\Delta H/\Delta t$ (j/g.s)	0.258	0.415	0.524	0.342	0.242	0.205

Moreover, nanofibers with better initial molecular orientation must be employed to prevent fiber shrinkage during thermal treatment. It is due to relax stain acquired by stretching during spinning at over Tg temperature and also, chemical reaction at high temperature which can disturb orientation of the molecular chains and causes poor mechanical properties of final fiber [14, 15].

Hence, the take up speed of 59.5 m/min by collecting nanofibers with better molecular orientation, low released heat caused by chemical reaction and suitable strength can be chosen as a desired speed for thermal treatment of PAN nanofibers. Work studying the effects of thermal treatment and stretching of electrospun PAN nanofiber simultaneously at elevated temperatures on the physical and mechanical properties is currently ongoing.

SUMMARY

A simple and non-conventional electrospinning technique using two syringe pumps was employed for producing of highly oriented PAN multifilament nanofibers. Process was carried out using two needles in opposite positions and a rotating collector perpendicular to needle axis. Current procedure was optimized for increasing of orientation and productivity of nanofibers with diameters in the nanoscale range. PAN nanofibers were electrospun from 14 wt% solutions of PAN in DMF at 11 kv on a rotating drum with various linear speeds from 22.5 m/min to 67.7 m/min. The influence of take up velocity was also investigated on the degree of alignment, internal structure, and mechanical properties of collected PAN nanofibers. Various characterization techniques were employed to find out the influence of operational parameters on degree of orientation. Based on micrographs captured by Optical microscope the APS was generated based on an image processing technique. The best degree of alignment was obtained for those nanofibers collected at take up velocity of 59.5 m/min. Moreover, polarized Raman spectroscopy under VV configuration at $\psi = 0°$ (parallel) and $90°$ (Vertical) versus polarization plane and also in VH configuration at $\psi = 0$ was used as a new technique for measuring molecular orientation of PAN nanofibers. Similarly, a maximum chain orientation parameter of 0.25 was determined for nanofibers collected at take up velocity of 59.5 m/min.

KEYWORDS

- **Crystallization index**
- **Electrospinning apparatus**
- **Polyacrylonitrile nanofibers**
- **Raman spectra**

Chapter 2

Development of Flexible Electrode Using Inkjet Printing of Silver Nanoparticles

Zahra Abadi, Vahid Mottaghitalab, Seyed M. Bidoki, and Ali Benvidi

INTRODUCTION

Biosensors represent a new trend emerging in terms of technological and theoretical achievements for the development and exploitation of analytical devices for detection, quantification, and monitoring of specific chemical species for clinical, environmental, and industrial analysis. In these devices the detection of a bio-chemical species is achieved by means of a biological element, through a traditional electronic transducer. In order to measure physiological parameters of complex environmental matrices *in vitro* and *in vivo* the miniaturization of these sensors is needed [1]. In biosensor the biological substances are used as recognition elements, which can convert the analytic substances concentration with the measurable electrical signals [2]. Metallic nanoparticles, a classic example of nanostructured materials, can be strongly considered for biosensor applications. Because of the ability to tailor the properties of nanomaterials, their incorporation into biosensors offers a great prospect of enhancing the performance of enzyme-based biocatalytic sensors [3]. They frequently display unusual physical and chemical properties, depending on their size, shape, and stabilizing agents. As for the catalytic and electronic properties of the metal nanoparticles, their application in electrochemistry, electron-transfer, and electrochemiluminescent (ECL) reactions are intensively investigated [4]. Especially for Pt nanoparticle [5, 6] and Au nanoparticle [7–13], they are widely used in modifying electrode.

It is well known that silver (Ag) is the best conductor among metals, and so silver nanoparticles (AgNPs) may facilitate more efficient electron transfer than gold nanoparticles (AuNPs) in biosensors [14]. Compared with gold, silver not only has good conductivity but also has antibacterial property. Furthermore, AgNPs are much cheaper than AuNPs and easy to be prepared, and have great potential applications in biosensor [15]. For example, Ren et al. [14] used Ag-Au nanoparticles to enhance the sensitivity of the glucose biosensors. They found that these Ag-Au particles could significantly enhance the current sensitivity of GOD enzyme electrodes. Lin et al. [16] reported the one-step synthesis of AgNPs/CNTs/chitosan film (Ag/CNT/Ch) hybrid film as a new alternative for the immobilization of GOD and HRP based on layer by layer technique.

There are many well-established manufacturing techniques for the fabrication of biosensors such as photolithography, screen printing, nano-imprinting, micro-contact printing, and dip-pen lithography. More recently, inkjet printing (IJP) is rapidly being developed for the display industry, the semiconductor industry (PCB, resister), and bio-medical technology array, sensor [17]. This technique offers benefits such as noncontact, speed, flexibility, creativity, cleanliness, competitiveness, and eco friendliness. IJP is an attractive method for patterning and fabricating objects directly from design or image files without the need for masks, patterns, or dies. In principle, IJP technology appears to be the simplest printing method, but it demands multi-disciplinary skills to precisely control the solid/liquid/gas interface [18].

However, the employment of IJP can solve many of the problems in a facile and effective manner. The IJP method allows for the patterning of conductive traces onto a substrate in one step, therefore reducing the time, cost, and space consumed and the toxic waste created during the manufacturing process [19].

In this chapter, the possibility of printing of AgNPs to prepare biosensor electrode using the conventional drop-on-demand mechanism demonstrated by inkjet technology is investigated. It was presumed that the IJP method allows for the patterning of conductive electrode onto a substrate in on step, therefore reducing the time and cost.

EXPERIMENTAL

Reagents
Ascorbic acid and silver nitrate were used both as analytical grade chemicals (99.5% purity, Merck). Double-distilled water used in all of the experiments without any further purification. Furthermore, a buffer solution was prepared from phosphate buffered saline (PBS, Sigma). An aqueous stock standard solution containing 0.01M H_2O_2 was freshly prepared for daily use. KNO_3 and H_2O_2 (35 wt%) were purchased from Merck company and used without pretreatment. Different substrates were used for the IJP experiments including A4 copying paper (80 GSM), plain weave fabrics from local supplier (Yazd Baft Co, Yazd, Iran) composed of (a) 100% cotton (210 g/m^2) (b) 100% polyester (PET) (170 g/m^2) (c) Cotton-PET(60/40) (175 g/m^2) (from Yazd-Baft, Yazd, Iran).

Instruments
A Hewlett Packard (*hp*) single head office inkjet printer (Apollo 1200) using color (*hp*25) and black (*hp*26) cartridges was used to print metallic salt and/or reducing agent solutions in separate runs. Usually, the substrate was first printed with the reducing solution and then with metal precursor ink. Microsoft Word was employed as the printer controlling software. A four-point probe method was used for measuring the electrical conductivity [20]. Cyclic voltammetric measurements were performed using an electrochemical analyzer CHI 660 (CH Instruments, Austin, TX, USA) connected to a personal computer. A three-electrode configuration was employed, consisting of an IJP silver electrode (4 mm^2) serving as a working electrode, whereas Ag/AgCl (3M KCl) and platinum electrode respectively served as the reference and counter electrode. Potentiometric experiments were performed using a metrohm pH meter (691). All potentials were reported with respect to the reference electrode. All electrochemical experiments were carried out at room temperature. Scanning electron microscopy (SEM) was run with an XL30 scanning electron micro analyzer (Philips, Netherlands)

at an acceleration voltage of 15 kV. The samples were coated with thin Au film towel characterize the film.

Preparation of Inkjet Printing Silver Electrode

A 30% w/v solution of ascorbic acid (pH = 5.5) and a 50.25% w/v silver nitrate solution (pH = 3.5) were used in the inkjet silver deposition electrode. The reducing ink normally should be printed first and after drying at room temperature for five minutes the silver nitrate solution is overprinted. Metal formation occurs *in situ* as a consequence of the redox reaction between reducing agent and metal salt solutions, as a consequence metallic layers composed of aggregated metal particles deposits over substrate. The thickness of the deposited layer may be increased by repeating the ink-jet deposition cycle. To extract the unreacted chemicals and their residues trapped among the deposited metallic particles during the redox reaction, the silver deposited pattern could be rinsed with water (wet extraction) and then hot pressed against a clean sheet of paper at 150°C (hot extraction). Some yellow residual compounds, extracted from the printed pattern, are transferred to the water and clean paper during rinsing and hot pressing procedures respectively.

Printed patterns (0.2 cm × 1.5 cm) then were glued to a polyester film for better handling during the electrochemical tests using PVC paste. Deposited electrodes then were cut and coated with water proof agent (Bit-guard Fc) with one end kept out of paste to allow for enough contact between the testing solution and the printed surface.

RESULTS AND DISCUSSION

Morphology of Inkjet Printed Layers of Agnps

The growth of deposited silver particle on paper substrate at different printing sequences was studied by considering letters A and G in the deposition sequence which respectively stands for IJP of ascorbic acid (A) and silver nitrate (G) inks. Table 2.1 lists the conductivity different pattern printed with various sequences. The best IJP sequence on paper was found to be AAAAGG which means printing four layers of ascorbic acid ink followed by two layers of silver nitrate solution according to the conductivity of 5.54×10^5 S/m. Figure 2.1 shows the SEM images of grown silver particles at different number of printed ascorbic acid layers.

The deposition of the silver nitrate (G) after printing of one layer of ascorbic acid (A) forms a non-homogenous aggregation of semispherical nanoparticle (Fig. 2.1-a). The morphology of printed layers of silver nitrate shows no significant change after printing of a double layer of ascorbic acid on paper substrate (Fig. 2.1-b).The IJP of silver nitrate based on AAAG (Fig. 2.1-c) and AAAAG (Fig. 2.1-d) protocols demonstrate a superior distribution of G nanoparticles on continues matrix of (A) fully spread on paper substrate. The change in morphology of inkjet printed layer occurs while the concentration of reducing agent (A) reaches to appropriate level. IJP of another layer of silver nitrate on AAAAG emerges a highly connected form of well defined semispherical AgNPs which is deposited on the paper surface (Fig. 2.2). The diameter of these semi-spherical nanoparticles is in the range of 80–200 nm.

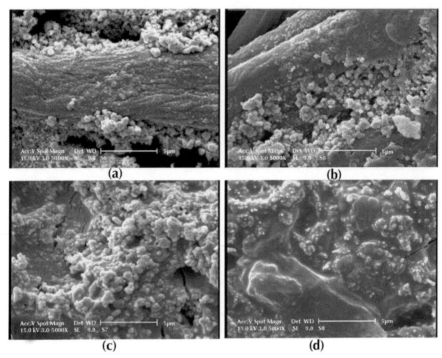

Figure 2.1. SEM images of quality growth of silver particles at different number of printed ascorbic acid layers (a) AG (b) AAG (c) AAAG (d) AAAAG.

Table 2.1. The conductivity of different pattern printed by various sequences.

Sample	Printing Consequence	Resistance (Ω)	Width (mm)	Length (mm)	Thickness (μm)	Conductivity ($\times 10^5 Sm^{-1}$)
1	AG	724.66	8.5	11	5	0.034
2	AAG	3.33	8.5	11	5	0.76
3	AAAG	1.89	8.5	11	5.5	1.24
4	AAAAG	1.64	8.5	11	6.5	1.20
5	AAAAAG	1.32	8.5	11	6.5	1.50
6	AAAGG	1.52	8.5	11	6	1.42
7	AAAAGG	0.34	8.5	11	7	5.54

Rinsing the printed surfaces with distilled water could also lower the final resistivity by extracting the reactants residue left on the surface and between the silver particles formed by the redox reaction although its effect was not in all cases as significant as the effect of heat treatment process. Removal of these impurities has been proved by SEM observations. Typical morphology of a deposited pattern is shown in magnified SEM images taken after hot extraction and rinsing procedures (Fig. 2.3).

Figure 2.2. SEM images of AgNPs electrodes on paper surface with AAAAGG sequence.

(a)   (b)

Figure 2.3. SEM images of a typical silver deposited layer on paper (a) Without pre treatment (b) Hot pressed at 150oC for 20 sec and rinsed.

Moreover, the Inkjet silver deposited rectangles was prepared on textile fabrics. Surface modification like coating the Cotton/Polyester (Co/PET) fabric using hydrophilic polymers such as Polyvinyl alcohol (PVA, 12% W/v solution) resulted in considerable improvement in overall conductivity as it could make a smooth ink receiving layer over the irregular surface of fabric. The best IJP sequence on PVA Coated Cotton/Polyester fabric was found to be AAAAGGAAAAGGAAAAGG with average conductivity of 1.4×10^5 S/m. The SEM images of Inkjet silver deposited on fabric surface as shown in Fig. 2.4. The diameter of these semi-spherical nanoparticles fell in the range of 100–900 nm.

(a)   (b)

Figure 2.4. SEM images of AuNPs nanoparticles on fabric surface in (a) low (b) high magnification.

Evaluation of Ag⁺ Ions by Direct Potentiometric Measurement

Potentiometric measurement was carried out on Ag^+ solutions with various concentrations (0.01, 0.005, 0.001, 0.0005, 0.0001, and 0.00005 M) using a two electrode cell including working electrode and a saturated calomel electrode (SCE). Potentiometry shows the performance of AgNPs compared to standard Ag working electrode (Fig. 2.5). The reference electrode was placed in $AgNO_3$ solution with various concentrations. A salting bridge containing KNO_3 was used to connect these two solutions. These results clearly indicated that the electrochemical behavior of electrodes was very similar.

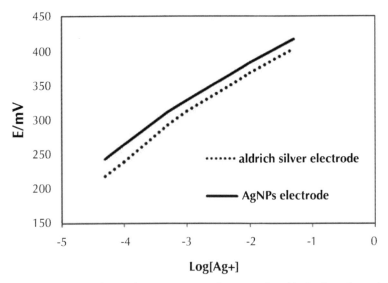

Figure 2.5. Plot of potential toward concentrations of Ag+ ions for Aldrich silver electrode (dashed line) and AgNPs electrode (solid line).

Electrochemical Response of Hydrogen Peroxide

Indication of Diabetic disease through Glucose measurement needs highly stable and sensitive biosensors. Glucose Oxidaze is commonly used to fulfill the basic requirement through enzyme immobilization. Hydrogen peroxide is direct product of enzymatic reaction of glucose in presence of glucose oxidaze (Scheme 2.1). The formation of Hydrogen peroxide releases two electrons which pass working electrode for producing electrical current.

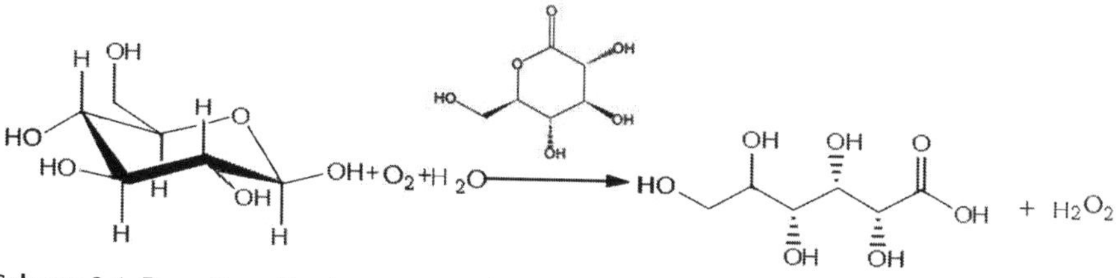

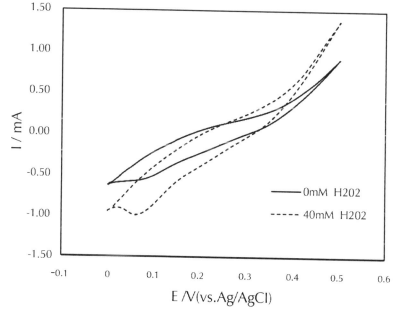

Scheme 2.1. Formation of hydrogen peroxide and Gluconic acid from enzymatic reaction of Glucose in presence of Glucose Oxidaze.

Figure 2.6 demonstrate the quantification of glucose through electrochemical detection of the enzymatically liberated H_2O_2 [21] in 0.1 M phosphate buffer solution (pH 7.0) in the potential window of $-0.4V$ to $+0.65V$ versus the printed Ag/AgCl reference electrode. The more hydrogen peroxide librates, the higher peak enhances in redox voltammetry. There was a substantial difference between the current for a 0 mM H_2O_2 solution (containing only the phosphate buffer) and the current for 0.4 mM and 0.8 mM H_2O_2 solutions. The results also indicate that inkjet AgNPs/paper electrode act as an effective factor for reduction of hydrogen peroxide, at relatively low potentials.

Figure 2.6. Cyclic voltammograms of the AgNPs/ paper electrode in a 0.1M phosphate buffer solution (pH 7.0) in a) 0 mM H_2O_2 b) 40 mM and c) 80 mM of hydrogen peroxide at 20 mVS^{-1}.

Similar to the inkjet AgNPs/paper electrode was compared to inkjet AgNPs/PVA coated Co/PET fabric electrode as the working electrode. As expected, weaker oxidation and reduction peaks were observed using fabric electrode with and without H_2O_2 solution compared to cyclic voltammogram for an inkjet AgNPs/paper electrode. Moreover, the electrochemical responses are markedly enhances at the AgNPs/paper electrode. This may be ascribed to the high specific surface and small size (80–200 nm) of silver particles at AgNPs/paper electrode (see Fig. 2.7).

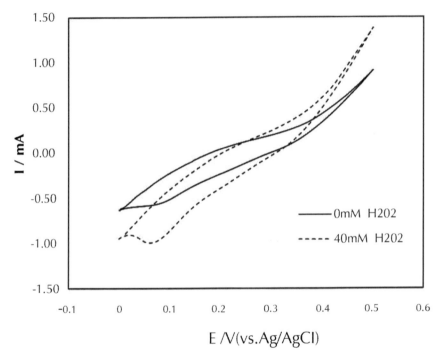

Figure 2.7. Cyclic voltammograms of the AgNPs/PVA coated Co/PET fabric electrode in a 0.1 M phosphate buffer solution (pH 7.0) in 0 mM H_2O_2 (solid line) and 40 mM (dashed line) of hydrogen peroxide at 20 mVS⁻¹.

Optimization of Experimental Condition

The pH dependence of the biosensor response was investigated over the pH range from 3 to 10, (Fig. 2.8). The anodic peak current increased monotonically with increasing pH value from 3.0 to 7.0 and then decreased with similar trend. Obviously, the maximum current response occurs at pH 7.0 therefore buffer solution of pH 7.0 was selected as the supporting electrolyte.

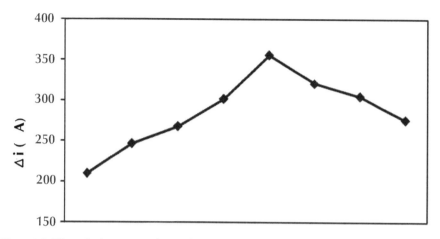

Figure 2.8. Effect of solution pH value on the oxidation peak current of AgNPs electrode in PBS (pH 3.0–10.0) containing 40 mM hydrogen peroxide at 20 mVS^{-1}.

Conductivity and Current

The relationship between anodic peak current and conductivity of AgNPs electrode was investigated using cyclic voltammetry. In this experiment, the AgNPs electrodes (0.5 cm^2) with different conductivity were examined. Figure 2.9 shows the results for the effect of conductivity using cyclic voltammetry in 0.1 M phosphate buffer solution (pH 7.0). The current is linearly proportional to the conductivity of AgNPs electrodes ($R^2 = 0.998$). This suggests that the whole process of the fabrication of electrode is a typical surface-controlled process, as expected for IJP systems.

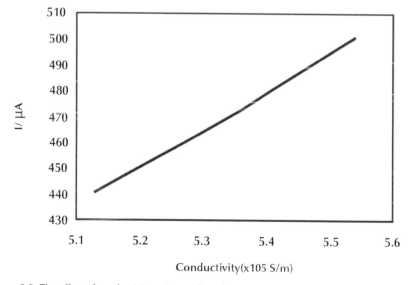

Figure 2.9. The effect of conductivity using cyclic voltammetry.

SUMMARY

A novel electrode for early determination of glucose was constructed. The AgNPs provides a good biocompatible environment and largely promote the direct electron transfer. The direct electron transfer was realized on the AgNPs electrode. The electrode showed effective activity toward the direct reduction of hydrogen proxide. The electrode also possesses a variety of good electrochemical characteristics that can be due to high special surface and size of silver particles. The resulting nanoparticles retained a good electrochemical response to hydrogen peroxide, which offered an excellent possibility of IJP AgNPs electrode for glucose biosensing.

KEYWORDS

- **Biosensors**
- **Cyclic voltammetric measurements**
- **Glucose oxidaze**
- **Inkjet printing**
- **Scanning electron microscopy**

Chapter 3

Supreme EMI Shielding Using Electroless Plating of Metallic Nanoparticle on Cotton Fabric

A. Afzali, V. Mottaghitalab, and M. Saberi

INTRODUCTION

Because of the high conductivity of copper, electroless copper plating is currently used to manufacture conductive fabrics with high shielding effectiveness (SE). It can be performed at any step of the textile production, such as yarn, stock, fabric or clothing [1].

Electroless copper plating as a non-electrolytic method of deposition from solution on fabrics has been studied by some researchers [1–9]. The early reported copper electroless deposition method uses a catalytic redox reaction between metal ions and dissolved reduction agent of formaldehyde at high temperature and alkaline medium [1, 2]. Despite of technique advantages, such as low cost, excellent conductivity, easy formation of a continuous and uniform coating, experimental safety risks appear through formation of hazardous gaseous product during plating process specially for industrial scale.

Further research has been conducted to substitute formaldehyde with other reducing agents coupled with oxidation accelerator such as sodium hypophosphite and nickel sulphate [4–8]. Incorporation of nickel and phosphorus particles providing good potential for creation of fabrics with a metallic appearance and good handling characteristics. These properties are practically viable If plating process followed by finishing process in optimized pH and in presence of ferrocyanide. Revealing the performance of electroless plating of Cu-Ni-P alloy on cotton fabrics is an essential research area in textile finishing processing and for technological design [4–9].

The main aim of this chapter is to explore the possibility of applying electroless plating of Cu-Ni-P alloy onto cotton fabric to obtain highest level of conductivity, washing and abrasion fastness, room condition durability and EMI SE. The fabrication and properties of Cu-Ni-P alloy plated cotton fabric are investigated in accordance with standard testing methods.

EXPERIMENTAL

Cotton fabrics (53×48 count/cm^2, 140 g/m^2, taffeta fabric) were used as substrate. The surface area of each specimen is 100 cm^2. The electroless copper plating process was conducted by multistep processes: pre-cleaning, sensitisation, activation, electroless Cu-Ni-P alloy deposition and post-treatment.

The fabric specimens (10 cm × 10 cm) were cleaned with non-ionic detergent (0.5 g/l) and NaHCO$_3$ (0.5 g/l) solution for 10 min prior to use. The samples then were rinsed in distilled water. Surface sensitization was conducted by immersion of the samples into an aqueous solution containing SnCl$_2$ and HCl. The specimens were then rinsed in deionized water and activated through immersion in an activator containing PdCl$_2$ and HCl. The substrates were then rinsed in a large volume of deionized water for 10 min to prevent contamination the plating bath. The electroless plating process carried out immediately after activation. Then all samples immersed in the electroless bath containing copper sulfate, nickel sulfate, sodium hypophosphite, sodium citrate, boric acid and potassium ferrocyanide. In the post-treatment stage, the Cu-Ni-P-plated cotton fabric samples were rinsed with deionized water, ethyl alcohol at home temperature for 20 min immediately after the metalizing reaction of electroless Cu-Ni-P plating. Then the plated sample dried in oven at 70°C.

The weights (g) of fabric specimens with the size of 100 mm × 100 mm square before and after treatment were measured by a weight meter (HR200, AND Ltd., Japan). The percentage for the weight change of the fabric is calculated in equation (1).

$$I_W = \frac{W_f - W_0}{W_0} \times 100\% \tag{1}$$

where I_w is the percentage of increased weight, W_f is the final weight after treatment, W_0 is the original weight.

The thickness of fabric before and after treatment was measured by a fabric thickness tester (M034A, SDL Ltd., England) with a pressure of 10 g/cm². The percentage of thickness increment was calculated in accordance to equations (2).

$$T_I = \frac{T_F - T_0}{T_0} \times 100\% \tag{2}$$

where T_I is the percentage of thickness increment, T_f is the final thickness after treatment, T_o is the original thickness.

A Bending Meter (003B, SDL Ltd., England) was employed to measure the degree of bending of the fabric in both warp and weft directions. The flexural rigidity of fabric samples expressed in N cm is calculated in equation (3).

$$G = W \times C^3 \tag{3}$$

where G (N-cm) is the average flexural rigidity, W (N/cm²) is the fabric mass per unit area, C(cm) is the fabric bending length.

The dimensional changes of the fabrics were conducted to assess shrinkage in length for both warp and weft directions and tested with (M003A, SDL Ltd., England) accordance with standard testing method (BS EN 22313:1992). The degree of shrinkage in length expressed in percentage for both warp and weft directions were calculated according in accordance to equation (4).

$$D_c = \frac{D_f - D_0}{D_0} \times 100 \qquad (4)$$

where D_c is the average dimensional change of the treated swatch, D_0 is the original dimension, D_f is the final dimension after laundering.

Tensile properties and elongation at break were measured with standard testing method ISO 13934-1:1999 using a Micro 250 tensile tester.

Color changing under different application conditions for two standard testing methods, namely, (1) ISO 105-C06:1994 (color fastness to domestic and commercial laundering, (2) ISO 105-A02:1993 (color fastness to rubbing) were used for estimate.

Scanning electron microscope (SEM, XL30 PHILIPS) was used to characterize the surface morphology of deposits. WDX analysis (3PC, Microspec Ltd., USA) was used to exist metallic particles over surface Cu-Ni-P alloy plated cotton fabrics. The chemical composition of the deposits was determined using X-ray energy dispersive spectrum (EDS) analysis attached to the SEM.

The coaxial transmission line method as described in ASTM D 4935-99 was used to test the EMI SE of the conductive fabrics. The set-up consisted of a SE tester, which was connected to a spectrum analyzer. The frequency is scanned from 50 MHz to 2.7 GHz are taken in transmission. The attenuation under transmission was measured equivalent to the SE.

RESULTS AND DISCUSSION

Elemental Analysis

The composition of the deposits was investigated using X-ray EDS elemental analysis. The deposits consisted mainly of copper with small amounts of nickel and phosphorus. Table 3.1 shows the weight percent of all detected elemental analysis.

Table 3.1. Elemental analysis of electroless copper plated using hypophosphite and nickel ions.

Element	Copper	Nickel	Phosphorous
~ wt%	96.5	3	0.5

The nickel and phosphorus atoms in the copper lattice possibly increase the crystal defects in the deposit. Moreover, as non-conductor, phosphorus will make the electrical resistivity of the deposits higher than pure copper. Electroless plating of copper conductive layer on fabric surface employs hypophosphite ion to reduce copper ion to neural copper particle. However the reduction process extremely accelerates by addition of Ni^{2+}. Addition of Ni^{2+} also sediments tiny amount of nickel and phosphorus elements. Following formulations show the mechanism of copper electroless plating using hypophosphite.

$$2H_2PO_2^- + Ni^{2+} + 2OH^- \rightleftharpoons Ni^0 + 2H_2PO_3^- + H_2$$

$$2H_2PO_2^- + 2\,OH^- \xrightarrow[\text{surface}]{Ni} 2e^- + 2H_2PO_3^- + H_2$$

$$Cu^{2+} + 2e^- \rightleftharpoons Cu^0$$

$$Ni^0 + Cu^{2+} \rightleftharpoons Ni^{2+} + Cu^0$$

Fabric Weight and Thickness

The change in weight and thickness of the untreated cotton and Cu-Ni-P alloy plated cotton fabrics are shown in Table 3.1.

The results presented that the weight of chemically induced Cu-Ni-P-plated cotton fabric was heavier than the untreated one. The measured increased percentages of weight were 18.47% .This confirmed that Cu-Ni-P alloy had clung on the surface of cotton fabric effectively. In the case of thickness measurement, the cotton fabric exhibited a 5.7% increase after being subjected to metallization.

Fabric Bending Rigidity

Fabric bending rigidity is a fabric flexural behavior that is important for evaluating the handling of the fabric. The bending rigidity of the untreated cotton and Cu-Ni-P-plated cotton fabrics is shown in Table 3.2.

Table 3.2. Weight and thickness of the untreated and Cu-Ni-P-plated cotton fabrics.

Specimen (10 cm × 10 cm)	Weight (g)	Thickness (mm)
Untreated cotton	2.76	0.4378
Cu-Ni-P-plated cotton	3.72 (↑18.47 %)	0.696(↑5.7 %)

The results proved that the chemical plating solutions had reacted with the original fabrics during the entire process of both acid sensitization and alkaline plating treatment. After electroless Cu-Ni-P alloy plating, the increase in bending rigidity level of the Cu-Ni-P-plated cotton fabrics was estimated at 11.39% in warp direction and 30.95% in weft direction respectively. The result of bending indicated that the Cu-Ni-P-plated cotton fabrics became stiffer handle than the untreated cotton fabric.

Fabric Shrinkage

The results for the fabric Shrinkage of the untreated cotton and Cu-Ni-P-plated cotton fabrics are shown in Table 3.3.

Table 3.3. Bending rigidity of the untreated and Cu-Ni-P-plated cotton fabrics.

Specimen	Bending (N·cm)	
	warp	weft
Untreated cotton	1	0.51
Cu-Ni-P-plated cotton	1.17(↑11.39%)	0.66(↑30.95%)

The measured results demonstrated that the shrinkage level of the Cu-Ni-P-plated cotton fabric was reduced by 8% in warp direction and 13.3% in weft direction respectively.

After the Cu-Ni-P-plated, the copper particles could occupy the space between the fibers and hence more copper particles were adhered on the surface of fiber. Therefore, the surface friction in the yarns and fibers caused by the Cu-Ni-P particles could then be increased. When compared with the untreated cotton fabric, the Cu-Ni-P-plated cotton fabrics shown a stable structure.

FABRIC TENSILE STRENGTH AND ELONGATION

The tensile strength and elongation of cotton fabrics was enhanced by the electroless Cu-Ni-P alloy plating process as shown in Table 3.4.

Table 3.4. Dimensional change of the untreated and Cu-Ni-P-plated cotton fabrics.

Specimen	Shrinkage (%)	
	warp	weft
Untreated cotton	0	0
Cu-Ni-P-plated cotton	–8	–13.3

The metalized cotton fabrics had a higher breaking load with a 28.44% increase in warp direction and a 35.62% increase in weft direction than the untreated cotton fabric. This was due to the fact that more force was required to pull the additional metal-layer coating.

The results of elongation at break were 12.5% increase in warp direction and 7.8% increase in weft direction, indicating that the Cu-Ni-P-plated fabric encountered little change when compared with the untreated cotton fabric. This confirmed that with the metalizing treatment, the specimens plated with metal particles was demonstrated a higher frictional force of fibers. In addition, the deposited metal particles which developed a linkage force to hamper the movement caused by the applied load.

COLOR CHANGE ASSESSMENT

The results of evaluation of color change under different application conditions, washing, rubbing are shown in Table 3.5.

Table 3.5. Tensile strength and percentage of elongation at break load of the untreated and Cu-Ni-P-Fabrics plated cotton fabrics.

Specimen	Percentage of elongation (%)		Breaking load (N)	
	warp	weft	warp	weft
Untreated cotton	6.12	6.05	188.1	174.97
Cu-Ni-P-plated cotton	6.98(↑12.5%)	6.52(↑7.8%)	241.5(↑28.4%)	237.3(↑35.62%)

The results of the washing for the Cu-Ni-P-plated cotton fabric were grade 5 in color change. This confirms that the copper particles had good performance during washing. The result of the rubbing fastness is shown in Table 3.5. According to the test result, under dry rubbing condition, the degree of staining was recorded to be grade 4–5, and the wet rubbing fastness showed grade 3–4 in color change. This result showed that the dry rubbing fastness had a lower color change in comparison with the wet crocking fastness. In view of the overall results, the rubbing fastness of the Cu-Ni-P-plated cotton fabric was relatively good when compared with the commercial standard.

Table 3.6. Washing and rubbing fastness of the untreated and Cu-Ni-P Fabrics plated cotton fabrics.

Specimen	Washing	Rubbing	
		Dry	Wet
Cu-Ni-P-plated cotton	5	4–5	3–4

SURFACE MORPHOLOGY

Scanning electron microscopy (SEM) of the untreated and Cu-Ni-P-plated cotton fabric is shown in Fig. 3.1 with magnification of 250X. Microscopic evidents of copper coated fabrics shows the formation of evenness copper particles on fabric surface and structure.

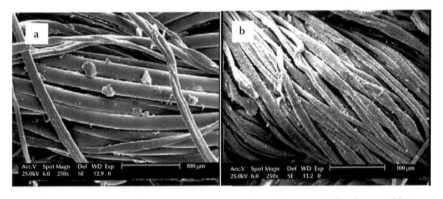

Figure 3.1. SEM photographs of the (a) untreated cotton fabric (b) Cu-Ni-P-plated cotton fabric.

Figure 3.2 shows the SEM and WDX analysis copper plated surfaces of cotton fiber. It was observed that the cotton fibers surface was covered by Cu-Ni-P alloy particles composing of an evenly distributed mass. In addition, WDX analysis indicated that the deposits became more compact, uniform and smoother also exist homogenous metal particle distribution over coated fabric surface. These results indicate that the effect of chemical copper plating is sufficient and effective to provide highly conductive surface applicable for EMI shielding use.

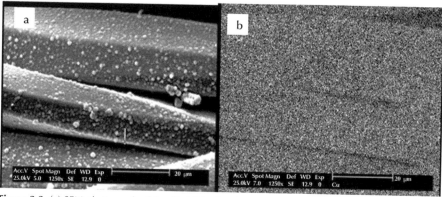

Figure 3.2. (a) SEM photograph of the Cu-Ni-P-plated cotton fabric (b) WDX analysis of the Cu-Ni-P-plated cotton fabric.

SHIELDING EFFECTIVENESS

Electromagnetic shielding means that the energy of electromagnetic radiation is attenuated by reflection or absorption of an electromagnetic shielding material, which is one of the effective methods to realize electromagnetic compatibility .The unit of EMI SE is given in decibels (dB). The EMI SE value was calculated from the ratio of the incident to transmitted power of the electromagnetic wave in the following equation:

$$SE = 10\log\left|\frac{P_1}{P_2}\right| = 20\left|\frac{E_1}{E_2}\right| \tag{5}$$

where P_1 (E_1) and P_2 (E_2) are the incident power (incident electric field) and the transmitted power (transmitted electric field), respectively. Figure 3.3 indicates the SE of the copper-coated fabrics with 1 ppm $K_4Fe(CN)_6$ compared to copper foil and other sample after washing and rubbing fastness test. The SE test applied on five different conductive sample including copper foil, electroless plated of Cu-Ni-P alloy particle on cotton fabric, electroless plated fabric after washing test, electroless plated fabric after dry and wet rubbing. As it can be expected copper foil with completely metallic structure shows the best SE performance according to higher conductivity compared to other conductive fabric sample. However, SE of copper-coated cotton fabric was above 90 dB and the tendency of SE kept similarity at the frequencies 50 MHz to 2.7 GHz. The acquired results for samples after washing fastness test shows nearly 10% decrease over frequency range which is still applicable for practical EMI shielding use. Two other samples after rubbing shows respectively 12% and 15% reduction in SE value but the presented results still show an accepted level of shielding around 80 dB. The SE reduction after fastness tests is a quite normal behavior which is likely due to removing of conductive particles from fabric surface. However, the compact and homogenous distribution of conductive particles provide a great conductive coating on fabric surface with high durability even after washing or rubbing tests. The copper-coated cotton fabric has a practical usage for many EMI shielding application requirements.

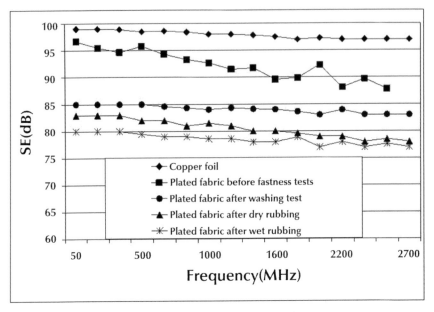

Figure 3.3. The shielding effectiveness of various conductive sample.

SUMMARY

In this study, Electroless plating of Cu-Ni-P alloy process onto cotton fabrics was demonstrated. Both uncoated and Cu-Ni-P alloy coated cotton fabrics were evaluated with measurement weight change, fabric thickness, bending rigidity, fabric shrinkage, tensile strength, percentage of elongation at break load and color change assessment. The results showed significant increase in weight and thickness of chemically plated cotton fabric. Coated samples showed better properties and stable structure with uniformly distributed metal particles. The SE of copper-coated cotton fabric was above 90 dB and the tendency of SE kept similarity at the frequencies 50 MHz to 2.7 GHz. Also, the evaluation of SE after standard washing and abrasion confirms the supreme durable shielding behavior. The copper-coated cotton fabric has a practical usage for many EMI shielding application requirements.

KEYWORDS

- **Bending meter**
- **Coaxial transmission line method**
- **Shielding effectiveness**
- **Surface sensitization**

Chapter 4

Inkjet Deposited Circuit Components

S. M. Bidoki, J. Nouri, and A. A. Heidari

INTRODUCTION

One of the ultimate goals in electronics is the ability to directly write electronic components and circuits. Recent scientific and technical articles are emphasizing the role of printing technology in the modern electronic industry for fabrication of items which has been named printed electronics. Advancing materials chemistry and developing print head technology is bringing this goal closer to reality. Printed electronics is a promising technology that has received tremendous interest as a mass production process for low-cost electronic devices, because it increases manufacturing flexibility and decreases manufacturing costs. The market for printed and potentially printed electronics, including organics, inorganics, and composites, is being forecasted to rise from $1.92 billion in 2009 to $57.16 billion in 2019. The low process temperatures allows the use of flexible substrate materials, such as paper or plastics, suitable for a reel-to-reel process [1].

Application of inkjet printing technique has been reported as a promising printing technique for manufacturing electronic components and devices. This technique offers benefits such as non-contact, speed, flexibility, creativity, cleanliness, competitiveness, and eco-friendliness. Inkjet printing is an attractive method for patterning and fabricating objects directly from design or image files without the need for masks, patterns, or dies. In principle, inkjet printing technology appears to be the simplest printing method, but it demands multi-disciplinary skills to precisely control the solid/liquid/gas interface [3].

Metallic particles dispersed in a polymeric matrix have been used conventionally as a paste or ink to print electrically active patterns on different substrates. In order to inkjet print with metals or ceramics, it is often necessary to print them as highly concentrated suspensions of powders in liquids. Such liquid suspensions must have physical properties appropriate to the inkjet delivery mechanism [4]. The potential of inkjet printing in this field is clearly important but the challenge to date has been how to achieve prints of low volume resistivity from the very low viscosity ink required for inkjet printing.

There are several approaches to formulate the metal precursor inks. They can be composed of dissolved organometallic compounds or polymers, colloidal suspensions of metal nanoparticles, or some combination of these constituents. This technology can achieve direct writing of patterns with feature sizes of 100 micron and conductivities of 50% and 75% of the bulk properties for silver and gold respectively. Post print metallization is still required which is a limitation and at present temperatures of

400°C are required and development work with the formulator to bring this down to 150°C is targeted [5].

In a recent method initially patented by QinetiQ in 2003, a process of depositing solid materials directly onto a substrate by using a fluid applicator such as an inkjet printer was introduced. According to this invention's concepts, a printed circuit board (PCB) could be printed by an inkjet printer by simply printing the metal salt and the reducing agent directly instead of two colors from a conventional inkjet printer [7].

Ascorbic acid as the reducing agent of choice in inkjet deposition of silver patterns on hydrophilic flexible substrate was the latest development. Aqueous solutions of silver nitrate and reducing agent were inkjet printed consecutively onto the substrate, where an immediate chemical reduction transforms the metal cations into very fine metallic particles. Printing a high quality pattern, appropriate for use in the electronic industry using this deposition technique needs more research as such findings in conjunction with the advanced capabilities of recently available inkjet printers has marked a starting point for a safer, low cost, and digitally controllable method for the build-up of metallic patterns on flexible substrates. Present report is the result of a part of our continuous researches on inkjet deposition techniques using this new approach toward single step fabrication of deposited circuit components for making different active and passive electronic circuits and devices.

EXPERIMENTAL

Chemicals
Ascorbic acid and silver nitrate were used both as analytical grade chemicals (99.5% purity, Merck). Different substrates were used in the inkjet metal deposition tests; these included A4 copying paper (80 GSM), A4 inkjet transparency film (polyester film with hydrophilic coating), and plain weaved fabrics made up of 100% cotton (210 g/m^2), 100% PET (170 g/m^2), and 60% Cotton/40% PET (175 g/m^2) which were obtained locally (from Yazd-Baf, Yazd, Iran).

Inkjet Deposition Plan
A Hewlett Packard (*hp*) single head office inkjet printer (Apollo 1200) using color (*hp*25) and black (*hp*26) cartridges was used to print metallic salt and/or reducing agent solutions in separate runs. Usually, the substrate was first printed with the reducing ink and then with metal precursor ink. Microsoft Word was used as the printer controlling software. The substrate could be an A4 paper, an inkjet transparency film or a piece of fabric glued to paper [8].

A 30% w/v solution of ascorbic acid (adjusted at pH = 5.5 by addition of NaOH solution) and a 50.25% w/v silver nitrate solution (pH = 3.5) were used in the inkjet silver deposition trials. Usually, the reducing ink was printed first and after an intermediate partial drying at room temperature for five minutes, the silver nitrate solution was overprinted. Metal formation occurred *in situ* as a consequence of the redox reaction between reducing agent and metal salt solutions leaving metallic layers composed of aggregated metal particles. The thickness of the deposited layer may be increased by repeating the inkjet deposition cycle.

To extract the unreacted chemicals and their residues trapped among the deposited metallic particles during the redox reaction, the silver deposited pattern could be rinsed with water (wet extraction) and then hot pressed against a clean sheet of paper at 150°C (hot extraction). Some yellow residual compounds, extracted from the printed pattern, are transferred to the water and clean paper during wet and hot pressing process respectively.

In order to inkjet deposit resistors/conductors on flexible substrates exhibiting required levels of electrical conductivity, solutions with different concentrations of ascorbic acid and silver nitrate were inkjet printed onto the substrates. Changing the sequence of printing layers of reducing and silver salt solutions could also build up silver patterns having divers packing density and conductivity.

A capacitor or condenser is a passive electronic component consisting of a pair of conductors separated by a dielectric. The simplest method of realizing an inkjet deposited capacitor is by printing two parallel conductive lines or rectangles on one side of a substrate. It is also possible to simply print two superimposed conductive lines or rectangles on both sides of the substrate using the substrate itself as the dielectric.

An inductor is a passive electronic component that stores energy in the form of a magnetic field. In its simplest form, an inductor consists of a wire loop or coil. Inkjet deposited inductors were printed on substrates simply by printing spiral or rectangular conductive tracks as coils or loops.

Measurements

A four-contact method was used for measuring the electrical conductivity of the deposited samples in preference to the four-point probe method usually employed in the electrical assessment of films [8, 9]. For each measurement, the inkjet metal deposited sample was placed over a glass surface and the four contact probes holder was positioned over the printed surface. The holder was fabricated with four copper probes which could contact the sample surface. To avoid any gap between probes and the printed surfaces, a weight of 500 g was placed over the probes holder to press them against the printed surface. The weight was exerting an optimum pressure over the printed surface which was chosen based on statistical assessments.

The conductivity of metal deposited samples with maximum length of 70 mm (minimum 55 mm) and of different print widths (maximum 30 mm) were measured; the actual length used in the resistance calculations was 10 mm which was the distance between the two inner probes of the four-contact device. After placing the sample between the glass plate and the probes holder, a constant current of 100 mA was supplied to the outer probes and the voltage between the inner probes was measured using a voltmeter. In this way an average value of resistivity over the width of the film strip (i.e., effectively the integration of an infinite number of four-point probe measurements across the strip width) could be obtained. To minimize measurement errors mainly caused by inconsistent contact between the copper probes and the printed surfaces, each sample was tested three times and the average value was used in conductivity calculations.

Inkjet silver deposited in rectangular shaped patterns was prepared on three kinds of substrates (paper, transparency film, and textile fabrics) and cut out for conductivity measurements. Each pattern was prepared using a different deposition sequence where each ink was ejected over the previously printed layer after an intermediate drying step. The conductivity was then measured and the average of three measurements on identically prepared samples was then reported as the final reading. For measuring the thickness of each deposited layer, the distribution map of elemental silver in the cross-sectional SEM image was used similar to what was explained in previous chapter [8].

A LCR400 Precision LCR Bridge (TTI Co., Bedfordshire, UK) was used to measure the impedance, inductance, and capacitance of the deposited components such as resistors, capacitors, and inductors.

RESULTS AND DISCUSSION

Inkjet Deposition of Conductors/Resistors

Inkjet printing of solutions with different concentrations of ascorbic acid and silver nitrate could lead to different levels of conductivity in deposited patterns. A 30% w/v solution of ascorbic acid (pH = 5.5) and a 50.25% w/v silver nitrate solution (pH = 3.5) found to posses the optimum concentrations of choice in the present experimental set up which could produce the highest degree of conductivity on all substrates. Higher concentration of the reactants in both inks was avoided as it could cause nozzle clogging especially in a thermal head inkjet printing set up.

Table 4.1. Tuning the level of conductivity by changing the printing sequence on paper as the printing substrate (A = jetting ascorbic acid ink, G = Jetting Silver nitrate ink).

Printing Sequence	Specific Resistivity (x10^{-5} Ωm)	Conductivity (×10^5 S/m)
AG	279.98	0.034
AAG	1.28	0.76
AAAG	0.8	1.24
AAAAG	0.82	1.20
AAAAAG	0.66	1.5
AGG	103.1	0.096
AAGG	1.11	0.88
AAAGG	0.7	1.42
AAAAGG	0.18	5.54
AAAAAGG	0.36	2.72
AGAG	1.25	0.78
AAGAG	0.71	1.42
AAGAAG	0.77	1.28
AAAGAG	0.44	2.24
AAAGAGA	0.34	2.94
AAAGAGAG	0.28	3.48

As the redox reaction between two inks is responsible for the *in situ* formation and build-up of conductive silver particles, changing the sequence of printing layers of inks could also build up silver patterns having divers packing density and degree of conductivity. It can simply change the quantity of each ink present on the substrate ready to react with the incoming droplets of the second reactant. Different printing sequences shown in Table 4.1, employed in inkjet deposition trials to evaluate the effect of ejecting dissimilar quantities of reactive inks on final conductivity of the deposited silver layer on paper.

Considering letters A and G in the deposition sequence which stands for inkjet printing of ascorbic acid (A) and silver nitrate (G) inks respectively, the best inkjet printing sequence on paper was found to be AAAAGG which means printing four layers of ascorbic acid ink followed by two layers of silver nitrate solution (marked in gray in Table 4.1). Taking into account the concentration of each ink and the output of each printing head (6.3 g/m^2 silver nitrate ink and 9.5 g/m^2 sodium ascorbate ink) reveals almost equal molar consumption of silver nitrate and ascorbic acid (1.31 respectively) in the above sequence for attaining the highest level of conductivity.

Heating the printed samples can improve the yield of the redox reaction by increasing the kinetic energy. Data presented in Table 4.2 is showing that the conductivity of inkjet deposited patterns could be improved significantly by heating the samples at 150°C for minimum 15 sec. Conductivity improvements presented in Table 4.2 is for samples chosen randomly which were produced using different inkjet printing sequences.

Rinsing the printed surfaces in distilled water could also lower the final resistivity by extracting the reactants residue left on the surface and between the silver particles formed in the redox reaction although its effect was not in all cases as significant as the effect of heat treatment process (Table 4.3).

Table 4.2. Changes in the conductivity ($\times 10^5$ S/m) of randomly chosen deposited samples (printed with different printing sequence) after heating at 150°C for different times.

Sample	Time 0 Sec.	5 Sec.	10 Sec.	15 Sec.	20 Sec.	25 Sec.
1	1.6	2.08	2.72	2.88	2.88	2.88
2	0.90	1.66	2.46	2.56	2.58	2.58
3	0.96	2.10	2.46	2.72	2.84	2.84
4	1.46	2.38	2.62	2.84	2.84	2.84
5	1.08	1.76	1.80	1.86	1.86	1.86
6	1.18	1.72	1.88	2.30	2.32	2.34
7	1.44	2.40	2.42	2.50	2.52	2.52
8	1.02	1.70	1.80	1.96	1.98	1.98
9	1.34	2.04	2.04	2.17	2.18	2.22
10	1.06	1.88	2.88	3.40	3.40	3.40

Table 4.3. Enhancement in the conductivity (x105 S/m) of randomly chosen deposited samples (printed with different printing sequence) on paper after rinsing.

Sample	1	2	3	4	5
Before Rinsing	1.96	2.18	3.02	1.44	1.24
After Rinsing	2.88	2.62	3.60	2.50	2.66

The main by-products of the redox reaction between silver salt and ascorbic acid are dehydroascorbic acid and sodium nitrate which are both soluble in water and can be washed away from the surface of the deposited particles by a simple rinsing procedure. Dehydroascorbic acid can also be softened at150°C which makes it sticky enough to be adhered to a clean cellulosic surface during the hot pressing stage against a clean sheet of paper. Removal of these impurities has been proved by SEM observations [8].

Flexibility is the main problem disturbing the efficient connection between deposited particles present in inkjet deposited layer. In a set of experiments, the inkjet printed patterns deposited on different substrates were folded to different angles and the changes in the overall conductivity were measured over the entire length of the samples (20 mm) using a normal multimeter (between two points). It was observed that the printed patterns deposited on paper and polyester (PET) film could bear deflections of up to 135 degree without any change in their overall conductivity while the printed patterns on textile fabric could only retain their conductivity for deflections of up to 100 degree which is mainly because of its texture and lower homogeneity of the silver deposited layer over its surface.

Table 4.4 demonstrates the highest level of conductivity obtained on different substrates and the printing sequence, each substrate needed to build up the required connection between the deposited particles based on its hydrophilicity and surface parameters. Coating the Cotton/Polyester (Co/PET) fabric with hydrophilic polymers such as polyvinyl alcohol (PVA, 12% w/v solution) resulted in considerable improvement in overall conductivity as it could make a smooth ink receiving layer over the fabric surface.

Table 4.4. Level of conductivity obtained on different substrates using an hp office inkjet printer.

Substrate	Printing Sequence	Conductivity ($\times 10^5$ S/m)
Paper	AAAAGG	5.54
PET film	AAAAGGAAAAGG	2.94
Co/PET Fabric	AAAAGGAAAAGGAAAAGG	0.168
Coated Co/PET Fabric	AAAAGGAAAAGGAAAAGG	1.4

PET transparent film itself has a smooth surface, but, because of its high surface tension, aqueous reactive inks could not spread over it to make a continuous and conductive thin layer of deposited particles over its entire surface. To resolve the above problem, a layer of ink receiving layer has been coated over its surface which in the present case has interfered with the redox reaction in such a way that the overall conductivity of the layers deposited on PET substrate was lower than layers deposited on paper.

Inkjet Deposition of Capacitors

A capacitor or condenser is a passive electronic component consisting of a pair of conductors separated by a dielectric. Several solid dielectrics are available, including paper, plastic, glass, mica, and ceramic materials. Paper was used extensively in older devices and offers relatively high voltage performance. The simple method of realizing a capacitor is by providing a slot in the middle of a conductive strip. Inkjet deposition of capacitors can be realized in the same manner by simply printing two conductive parallel lines or rectangles on one side or as superimposed patterns on two sides of a substrate.

Capacitance is a direct function of the cross-sectional area of the conductor. The area of interaction can be improved by incorporating a comb structured electrode in between the strip which is called inter digital structure or by lengthening the parallel lines in a spiral format. Inkjet deposition technique was used in a set of experiments to fabricate capacitors with sufficient capacity to be suitable for using in electronic circuits. Inter digital, spiral, and parallel plate capacitors were inkjet printed on paper as a typical ink receiving substrate to evaluate the performance of each capacitor. Deposited capacitors are demonstrated in Fig. 4.1 alongside with their capacitance measured on a LCR meter.

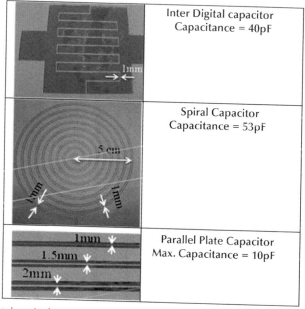

Figure 4.1. Inkjet deposited capacitors printed on paper with their maximum capacitance measured by a LCR meter.

It was observed that printing thinner parallel lines as long and close as possible could increase the final capacitance of the printed capacitor although the print quality would be very important in fabrication of such patterns. To compare the performance of the deposited capacitors with conventional copper etched items, a spiral capacitor identical to the inkjet deposited capacitor was etched on a copper board. The capacitance of the etched capacitor was measured which was 75 pF which was close to the obtained value from the deposited spiral capacitor (53 pF).

To fabricate capacitors with higher capacitance, two-sided capacitors (parallel plate capacitor) were prepared by gluing two piece of substrate each having a square shaped deposited pattern on them. It is also possible to fabricate a two-sided capacitor by simply printing on both sides of the substrate where the substrate's thickness is enough to avoid electrical current between conductive patterns. Two-sided capacitors made by gluing different substrates are shown in Fig. 4.2 alongside with their capacitance.

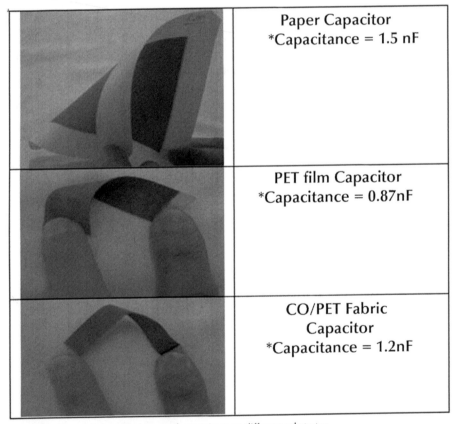

	Paper Capacitor ***Capacitance = 1.5 nF**
	PET film Capacitor ***Capacitance = 0.87nF**
	CO/PET Fabric Capacitor ***Capacitance = 1.2nF**

Figure 4.2. Two-sided inkjet deposited capacitors on different substrates.
*Capacitance reported belongs to a two-sided capacitor with 4.5 ꓴ 4.5 cm conductive plate on each side of the substrate.

Inkjet Deposition of Inductors

An inductor is a passive electronic component that stores energy in the form of a magnetic field. In its simplest form, an inductor consists of a wire loop or coil. The inductance is directly proportional to the number of turns in the coil. Inductance also depends on the radius of the coil and on the type of material around which the coil is wound.

Inkjet deposited inductors can be printed on substrates simply by printing spiral or rectangular conductive tracks. Spiral inductors shown in Fig. 4.3 were prepared on paper with radius of 5 cm composed of nine turns from perimeter to the center. Similar inductor was etched out of a copper board and the resulting inductance was compared between etched and inkjet deposited inductors. Based on our findings, the most important parameter in designing a useful deposited inductor is the overall end-to-end resistance of the inductor coil. Inkjet deposited tracks in the above mentioned format needed to have resistance of less than 100Ω to show acceptable level of inductance. The performance of the deposited and etched inductor is shown in Fig. 4.3 along with the end-to-end resistance of each coil.

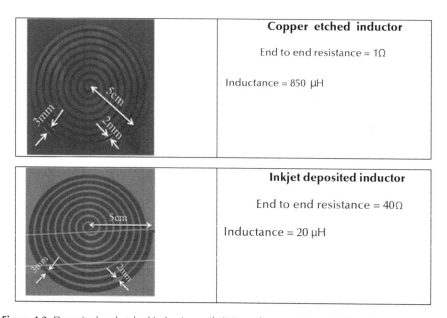

Figure 4.3. Deposited and etched inducting coils (10 cm diameter, 3 mm thickness).

Decreasing the end-to-end resistance of the deposited coil, as mentioned earlier, is achievable by printing thicker layers and wider tracks which lead to higher inductance of the inkjet deposited inductor. Although it is possible to build up thicker layers by repeating the inkjet printing sequence which can straightforwardly accomplish on industrial inkjet printers, depositing new layers exactly on top of the previously printed pattern using an office inkjet printer is very problematic and a hurdle for producing inductors with higher inductance.

SUMMARY

Inkjet deposition technology was used in the present research to fabricate different electronic components including resistors/conductors, capacitors, and inductors. Inkjet deposition procedures were performed on hydrophilic substrates such as paper, coated PET film, and textile fabrics using a thermal office inkjet printer.

Various levels of conductivity could be easily achieved by just changing the packing density of the deposited silver particles grown *in situ* from different mixtures of reducing and metal salt inks ejected onto the ink receiving layer. Substrate surface roughness is the main parameter which dictates the necessity of depositing a thicker layer on uneven surfaces for attaining an acceptable level of conductivity. The highest conductivity was achieved on inkjet deposited conductors fabricated on paper followed by on PET film and textile fabrics. Surface modification of the substrate prior to the inkjet depositing process, using hydrophilic polymers such as PVA, could improve the level of conductivity. Coated film works as an ink receiving layer itself which is smoother than the underlying substrate such as textile fabrics. It enhances efficient film formation by aqueous inks prior to particle deposition on the surface which improves the possibility of forming a continuous conductive layer.

Particle accumulation during the inkjet deposition process normally takes place when each individual droplet grows thousands of particles according to the concentration profile of the reducing agent and silver salt at each particular position. Ejection of almost equal molar quantity of silver salt and reducing agent was found to be necessary for attaining the highest level of conductivity on any substrate.

Post treatment of the inkjet deposited silver patterns using heat and water could also improve the ultimate level of conductivity on different substrates. The office thermal head inkjet printer employed in this research (300 dpi) could deposit silver patterns on substrates which (on paper) was almost 110 times lower conductive than bulk silver (6.3×10^7 S/m). It is presumable that the conductivity of deposited patterns can be enhanced further by using high resolution printers and smoother hydrophilic substrates.

Clearing the surface of the deposited pattern and completion of the redox reaction during the hot extraction process which involves application of heat and pressure could lead to improvement of electrical conductivity by increasing the compactness of the particles and lowering the tunneling resistance among them. Rinsing was also an effective method of lowering the patterns resistivity by dissolution and removals of non conductive reaction by-products such as dehydroascorbic acid and sodium nitrate.

Capacitors can also be fabricated using the inkjet deposition technique in different formats as parallel lines, plates, spiral, and inter digital shape. Although spiral shape deposited capacitor could perform like a normal circuit component, but higher capacitance could be achieved by fabricating parallel plate capacitors by printing conductive patterns on both sides of the substrate. Components with capacity in range of nano

Farad could be simply fabricated using the present inkjet deposition technique. In this configuration the substrate itself will work as the dielectric compound.

Induction coils were also deposited on paper as printing substrate which upon achieving very high level of conductivity they could perform as an efficient inductor capable of integration in different electronic circuits. Inkjet deposition of thicker and wider conductive tracks by repeating the deposition cycle, as many times as necessary, can reduce the end-to-end ohmic resistance of the deposited planar inductive coil which enhances its performance as a suitable inductor in different electronic devices.

Although printing high quality electronic components, appropriate for use in the electronic industry needs more research, the present findings in conjunction with the advanced capabilities of recently available inkjet printers mark a promising point for a safer, low cost, and digitally controllable method for the build-up of metallic patterns as circuit components. Acceptable performance of the deposited components along with the low consumption of energy and chemicals, digital control over the ejection and positioning process, user friendliness, and so forth are some of the advantages realized on replacing conventional production techniques with inkjet deposition procedures.

ACKNOWLEDGMENT

The authors would like to thank members of staff in the department of electrical and computer engineering at Yazd University for their kind involvement and scientific supports.

KEYWORDS

- Capacitor
- Dehydroascorbic acid
- Inductor
- Printed circuit board
- Redox reaction

Chapter 5

Reinforcing Chitosan/Poly(Vinyl alcohol) Nanofiber Scaffolds using Single-Walled Carbon Nanotube for Neural Tissue Engineering

Mohammad Ali Shokrgozar, Fatemeh Mottaghitalab,
Vahid Mottaghitalab, and Mehdi Farokhi

INTRODUCTION

Reconstruction of gray and white matter defects in the brain is especially difficult after large lesions or in the chronic situation, where an injury has occurred sometime previously [1–3]. In addition to neural death, there is demyelination, rejection or aberrant sprouting of injured axons, glial/minengeal scarring, and often progressive tissue cavitations. In these circumstances, there is a need for cell replacement and some form of neural tissue engineering to develop scaffolds that facilitate reconstruction and restore continuity across the traumatized region [4–9]. Among various scaffolds useful for sustained three-dimensional growth of neural cells, porous nanofiber composites have shown great potential to mimic the natural extracellular matrix (ECM) in terms of structure, porosity, and chemical composition. Thus, these nanofibers must be adequately processed to obtain a porous matrix of suitable morphology [10–12]. Electrospinning is the most effective method which has recently established the reputation for its capability to produce nanofiber composite scaffolds, and is counted as a new addition to the conventional techniques (e.g., phase separation, and self-assembly) [13–16]. In electrospinning process, a high voltage is applied to a polymer solution that is pumped to a spinneret facing an earthed target (collector). Upon reaching a critical voltage, the surface tension of the polymer at the spinning tip is counterbalanced by localized charges generated by the electrostatic force, and the droplet elongates and stretches into a Taylor cone where a continuous jet is ejected [17–20].

Electrospun nanofibers can be made of natural and synthetic materials with a combination of reinforcing agent to form composite nanomaterials [16, 21]. In the case of nanofiber-reinforced composite materials; a phase consisting of strong, stiff fibrous components is embedded in a more ductile matrix phase. Chitosan [β-(1,4)-2-amino-2-deoxy-D-glucose] is a unique polysaccharide obtained by deacetylation of chitin, a natural polymer having a primary amino groups at C_2 and hydroxyl groups at C_3 and C_6 positions [6, 22]. It is environmentally friendly, nontoxic, and biodegradable. To enhance the electrospinnability and structural properties of chitosan, blend systems with poly (vinyl alcohol) has been explored [17–19]. PVA is a synthetic polymer with non-toxic, non-carcinogenic, and bio-adhesive properties. Applying the blend system of CS/PVA would combine the properties that are unique to each such as superior structural properties and biocompatibility [22, 23]. However, the main disadvantage of the obtained nanofibers is low volume porosity which is not suitable for some neural tissue engineering applications [15, 24]. In the contrary, porous scaffolds has the ability to promote initial cell attachment and subsequently migration into and through the matrix, mass transfer of nutrients and metabolites, vascularization, and permit sufficient space for development and later remodeling of organized tissue [11, 13–15, 25]. For this purpose, single-walled carbon nanotubes (SWNTs) were used as an additive in CS/PVA fiber-reinforced composites due to their high aspect ratio, porosity, and chemical orientation [26, 27]. SWNTs are essentially pure carbon polymers, with each carbon atoms in the lattic bound covalently to only three neighboring carbon atoms. The highly symmetric beam-and truss structure formed by the covalently bound carbon atoms gives high stability and strength, while retaining flexibility [28, 29]. In the present study, chitosan/poly(vinyl alcohol) reinforced SWNT (SWNT-CS/PVA) nanofibrous membranes have been developed as scaffold. It is suggested that, the combination of SWNTs with CS/PVA blends may improve the morphology and porosity of the scaffold. For this purpose, a number of properties of the generated nanofibers such as molecular identity, porosity, and morphology are characterized. Moreover, the biocompatibility of nanofibrous scaffolds is also evaluated using human brain-derived cells and U373 cell lines.

MATERIALS AND METHODS

Materials

Materials used in this research included chitosan (powder, medium molecular weight, degree of deacetylation, DD% = 80%, Sigma-Aldrich, USA), PVA (degree of hydrolysis = 80%, degree of polymerization, approximately 2500, Sigma-Aldrich), SWNTs (TOP NANOSYS Inc.), Fibroblast growth factor (bFGF; Sigma-Aldrich), RPMI 1640 (Biowhittaker, Belgium), fetal bovine serum (FBS; Sigma-Aldrich, USA), Dulbecco Modified Eagle's Medium (DMEM; Biowhittaker, Belgium), penicillin (GIBCO), streptomycin (GIBCO), U373-MG cell line (Human glioblastoma-astrocytoma; National Cell Bank of Iran, NCBI), Scanning electron microscopy (SEM; Zeiss, Germany), MTT (Sigma-Aldrich, USA), Isopropanol (Sigma-Aldrich, USA).

Methods

Preparation of CS/PVA Reinforced SWNT Nanofibers using Electrospinning

To fabricate CS/PVA reinforced SWNT nanofibrous membranes with electrospinning process, chitosan was dissolved in aqueous acetic acid at room temperature with gentle stirring for two hours to form 2 wt% homogenous solutions. SWNTs were ultrasonicated in a 2% acetic acid aqueous solution with a sonic tip for two to five minutes to form highly dispersed and suspended particles. Ultrasonication was performed at a constant frequency of 25 KHz and the temperature was kept at 25°C throughout the deposition process. The prepared CS solution was added to SWNTs in a period of time with a magnetic stirrer agitation to make a black and concentrated solution. At the same time, PVA was dissolved in distilled water at 80°C with gentle stirring for 2 hr to form a 12 wt% homogenous solution. Then, the PVA solution was mixed with the SWNT reinforced CS solution. CS/PVA reinforced SWNT solution was prepared in

7%, 12%, and 17% concentrations of SWNT and CS/PVA ratio was remained constant at 25/75 ratio in all solutions. After air bubbles were removed completely, the mixed solution was placed in a plastic syringe bearing a 1-mm inner diameter metal needle which was connected with a high voltage power supply. The grounded counter electrode was connected to an aluminum foil collector. Electrospinning was performed at 15-KV voltage, 10 cm distance between the needle tip and the collector. It usually took 4 hr to obtain a sufficiently thick membrane that could be detached from the aluminum foil collector.

Extraction of Nanofibers
ISO 10993-5 standard was used to extract the samples for indirect biocompatibility tests [30]. Briefly, all nanofibrous matrices with the surface area of 1 cm^2 were soaked in 1 mL of culture medium in falcon tube and were incubated at 37°C for 3, 7, and 14 days for cell viability assays.

Cell Culture

U373-MG and human brain-derived cells originally isolated from human brain were used to evaluate the biocompatibility of generated nanofibers. U373-MG cell lines were cultured in RPMI 1640 with 10% fetal bovine serum (FBS; Sigma-Aldrich, USA). Human brain-derived cells were also cultured in Dulbecco Modified Eagle's Medium with 10% FBS, 100 Unit/ml penicillin, 100 μg/mL streptomycin 5 μg/mL L-glutamine, and 10 ng/mL fibroblast growth factor. Both cell types were incubated at 37°C with 5% CO_2. The cells were utilized between four and eight passages.

Characterization

Morphology Characterization
The nanofibers collected onto aluminum plates were cut into small pieces and were assessed using scanning electron microscopy. Moreover, the morphology of the cells that had been seeded (for seven days) on CS/PVA reinforced SWNT nanofiber composites was also evaluated using SEM. Briefly, the cellular constructs were harvested, washed in PBS, and then fixed in 4% paraformaldehyde for 15 min and dehydrated with graded concentration (50–100% v/v) of ethanol. Subsequently, they were kept in a hood for air drying. After sputter-coated with gold, samples were examined with scanning electron microscopy. The quality of SWNTs nanoparticles distribution in CS/PVA matrix was also evaluated using a transmission electron microscope (TEM) model CM200 (Philips).

Raman Spectroscopy
Raman images were taken with an Almega Thermo Nicolet Dispersive Raman Spectrometer using a short working distance 100 × objective and with resolution of 4 cm^{-1}. The laser source used was a second harmonic 532 nm of a Nd:YLF laser to avoid excessive fluorescence in the Raman signal.

Porosity Analysis

In order to measure pore characteristics of electrospun nanofibers, image processing method was performed. For this purpose, SEM micrographs of the nanofibers were required. The critical step in measuring pore characteristics is initial segmentation to generate binary images [31, 32]. In this approach, each image was divided into subimages where the heterogeneities were insignificant. The pore size is the diameter of a spherical particle that can pass the equivalent square opening, hence the pore size, O_i, results from:

$$O_i = (A_i)^{1/2} \qquad (1)$$

Thus, the pore size distribution (PSD) could be calculated from the area of pores and porosity. The PSD curve can be used to determine the uniformity, Cu, of the investigated materials, the uniformity coefficient is a measure for the uniformity of the openings and is given by:

$$Cu = O_{60}/O_{10} \qquad (2)$$

In this case, the data in pixels could be converted to nm. Finally, a PSD curve was plotted and O_{50}, O_{95}, and Cu were determined.

Mechanical Properties

The tensile strength of the samples was determined using a uniaxial tensile testing device equipped with a 50 N load cells and the crosshead speed was 5 mm/s. The average value of the stress at break and elongation at break was set as the representative value. Each result was taken from three (replicates, $n = 4$) rectangular specimens with dimension of 60 mm 4 mm (length width), according to ASTM D882-02 (2002) standard, cut from the films obtained by electrospinning. The thickness of films produced were (100 $\pm 25\mu m$) depending on the blend composition, resulting on sample strips with average cross-section area of 0.3–0.5 mm².

Cell Viability Assessment

MTT Assay

The cell viability was evaluated using 3-[4, 5-dimethyltriazol-2-y1]-2, 5-diphenyl tetrazolium bromide (MTT) as a substrate. Tetrazolium salts are reduced to formazan by mitochondrial succinate dehydrogenase, an enzyme which is active only in cells with an intact metabolism and respiratory chain. Firstly, 2×10^4 cells were seeded on matrices within a 96-well plate. The cells were incubated at 37°C and 5% CO_2. After 24 hr incubation, the culture mediums was replaced with 100 µl/well extraction solution of nanofibrous membranes and were incubated at 37°C and 5% CO_2 for additional 24 hours. Finally, the extraction solution of membranes was removed and was replaced with 100 µl/well MTT solution and re-incubated for four hours. After that, the formazan reaction products were dissolved in isopropanol solution for 15 min. The optical density of the formazan solution was read on an ELIZA plate reader at 570 nm. Moreover, TPS (Tissue Culture Polystyrene) and CS/PVA nanofiber composites (composite control) were used as control groups.

Neutral Red Assay (NR)

NR is taken up into the lysosomes by viable cells. All stages were identical to the stages mentioned in MTT assay instead of the third stage; after removing the extraction solution of nanofibrous membranes, 100 μL/well NR solutions were added each wells. After four hours, stain extraction solution (1% glacial acetic acid, 50% ethanol, and 49% distilled water) was replaced with NR solution and was re-incubated in 37°C and 5% CO_2 for 15 min. The absorbance was measured at 570 nm using ELISA reader.

Statistical Analysis

One-way analysis of variance (ANOVA) followed by post hoc Dunnet test was used to determine the statistical differences between the experimental groups (SPSS 16.0, USA). The values were considered significantly different if the p-value was <0.05.

RESULTS AND DISCUSSION

Electrospun CS/PVA reinforced SWNT nanofiber mats

Recently, there has been great interest in the biological applications of SWNTs in neural tissue engineering applications [6, 10, 27]. SWNTs have the potential to provide the needed structural reinforcement for neural tissue scaffolding due to their high aspect ratio (>1000), porosity, and high structural and chemical stability [33–37]. In the present study, to provide multifunctional structural reinforcement needed for newly created neural tissue scaffold, SWNTs have been put into a host of CS and PVA nanofiber composites [38–43]. The electrospun chitosan/PVA nanofibers easily forms hydrogel structure in an aqueous solution such as cell culture media [44]. However, based on experimental observation the addition of SWNTs improves the mechanical stability of composite nanofiber scaffolds in aqueous media. In fact, the SWNTs promote the rigidity of nanofiber and prohibit its tendency to shrinkage [45, 46]. Combination of SWNTs with CS/PVA to form hybrid nanofiber composites may represent a new avenue to take advantage of SWNTs properties. One of the major problems in SWNTs composite research is to obtain a suitable dispersion and distribution of the filler. The individualization of SWNTs from primary agglomerates is very difficult due to their high Vanderwaals forces and physical entanglement [21, 27]. To overcome this problem in the current study, ultrasonication was used to fabricate highly dispersed CS/SWNT nanocomposites. Ultrasonicating can cause quick and complete degassing from solution, accelerated chemical reactions; improved diffusion rates; and highly uniform dispersions of particles [47–50]. Therefore, the highly dispersed SWNT/CS solution combined with PVA would be excellent building block to act as structural and functional analogue of the original tissue for neural tissue engineering applications.

Morphology of Fibrous Scaffolds and the Seeded Cells

Figure 5.1 represents SEM images of bulk CS/PVA composite with different SWNT contents.

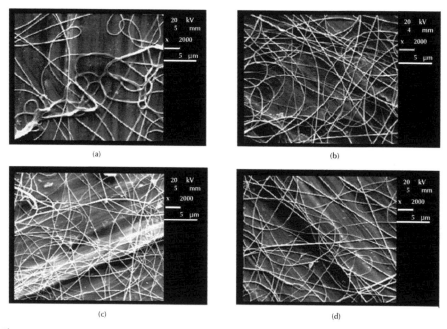

Figure 5.1. Scanning electron micrographs of: (a) CS/PVA nanofibers, (b) CS/PVA nanofibers with 7% SWNT, (c) CS/PVA nanofibers with 12% SWNT, (d) CS/PVA nanofibers with 17% SWNT. (Magnification: 2000X)

The acquired images show very similar morphological aspects for the CS/PVA samples at 25/75 ratios, with uniform and continuous entangled ribbon and or nanofiber with circular cross section. The higher the SWNT content, the more non-entangled structure with homogenous and circular cross section can be produced. It is suggested that polymers, CS and PVA, prior to SWNT addition have their chain mostly physically entangled in the hydrogel network. But presumably, a bonded hydrogel with higher electrical charge after SWNT addition has taken place. In fact, the SWNT nanoparticles incorporated in CS/PVA matrix may promote the electrical charge of bare CS/PVA [21, 33]. Therefore, the formation of physically entangled structure is prohibited. Further, characterization carried out to investigate the distribution of the SWNTs in SWNT-CS/PVA nanofibers. Figure 5.2 shows a fairly homogeneous dispersion of SWNTs throughout the CS/PVA matrix based on TEM micrograph.

The SWNTs particles showed diameters between 10–30 nm. It can be reasonably expected that the black particles are cross section of SWNTs bundles that were wholly distributed in the matrix. Our hypothesis is that entrapment of SWNT in CS/PVA matrices will tailor the morphology and porosity of nanocomposites. Thus, as it has been shown in Fig. 5.3 (a, b), SWNT-CS/PVA scaffolds have provided excellent 3D synthetic environments to bring brain-derived cells and U373 cell line in close proximity so that they can assemble while retaining their normal morphology.

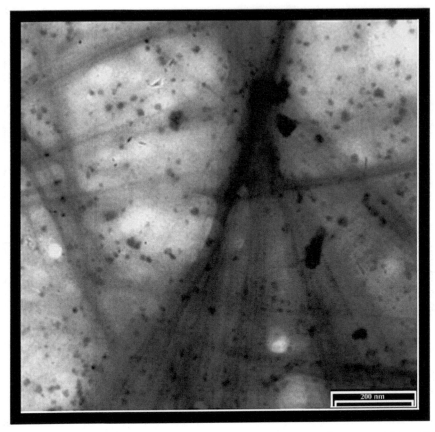

Figure 5.2. TEM micrographs of nanofiber thin film, showing SWNTs dispersion within a 17% SWNTsCS/PVA sample.

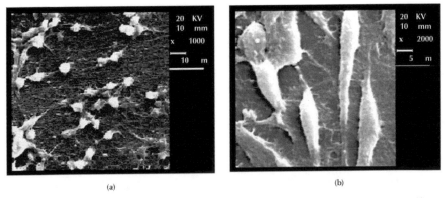

Figure 5.3. Scanning electron micrographs of the cells seeded on SWNT-CS/PVA nanofibrous membranes: (a) Brain-derived cells (Magnification: 1000X), (b) U373 cell lines. (Magnification: 2000X).

It is suggested that SWNTs could impart and accelerate the growth of cells in comparison to bulk CS/PVA nanocomposites. CS/PVA reinforced SWNT nanofibrous membranes form a supportive meshwork around cells and provide anchorage to the cells. In this well-defined architecture, the adhesion, proliferation, and migration of cells are influenced.

Raman Spectroscopy

Raman spectroscopy was used to evaluate changes in the molecular structure of CS/PVA nanocomposites prior and after SWNTs addition. SWNTs showed very unique peaks in Raman spectroscopy (Fig. 5.4). Four prominent sets of peaks which are typical of SWNTs were observed in the low- (~200 cm⁻¹, RBM [radial breathing mode]), moderate frequency (1600 cm⁻¹, G band, and 1300 cm⁻¹, D band) and high-frequency (2600 cm⁻¹, D* band) regions [33, 51, 52].

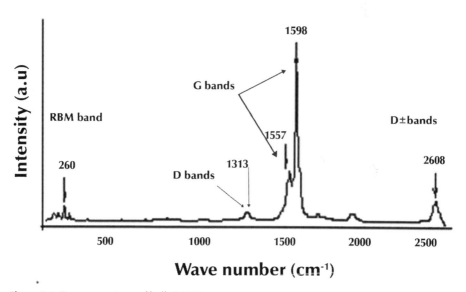

Figure 5.4. Raman spectrum of bulk SWNTs.

The Raman signature attributed to SWNT-CS/PVA nanofiber composites (Fig. 5.5a) shows a combination of enhanced spectra with significant Raman shift for spectra of individual CS/PVA [Fig. 5.5(b)] and bulk SWNT. As shown in Fig. 5.4, a comparison between bands enhanced for bare CS/PVA and those nanofibers including SWNTs indicates that the characteristic peaks of CS/PVA shifts to shorter wave numbers than by addition of SWNTs.

A shift to shorter wave numbers and hence lower frequency implies that the CS/PVA in the chemically interacted network requires relatively more energy to vibrate. In this case, the produced nanocomposites would have high aspect ratio with covalent bond between the reinforcement phase (SWNTs) and the matrix materials (CS/PVA) in an ideal state. The CS/PVA polymer fibers are structured in their local orientation to generate the desired mechanical properties by the addition of SWNT as a second phase.

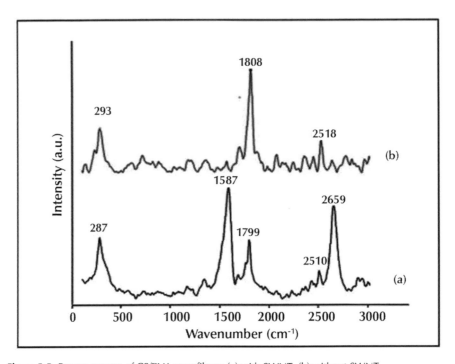

Figure 5.5. Raman spectra of CS/PVA nanofibers: (a) with SWNT, (b) without SWNT.

Porosity Analyzing

Figure 5.6 shows the PSD curves. The electrospun membrane based on CS/PVA containing 17% SWNT exhibited the highest porosity value (~73%), which was higher than the values determined from CS/PVA nanofibers containing 7% and 12% SWNT and about 20% more than the bulk CS/PVA nanofibers (~54%).

This can be attributed to promotion of electrical charge after addition of SWNTs to CS/PVA matrix. The presence of SWNTs charged particle increases the electrical conductivity of spinning solution and consequently improves the web arrangement through efficient orientation of nano-arrays. It is suggested that the increase of SWNT content in CS/PVA nanofibers leads to greater number of crossovers. Therefore, large pores are split into several smaller pores. As a result, O_{50} and O_{95} decrease [31]. Furthermore, this fracture of the pore results less variation of the pore size. Hence, uniformity increases. It is important to consider that most cells have sizes on the scale of microns and thus larger pore sizes are needed to allow for cellular infiltration into scaffolds. Thus, the highly porous architecture of CS/PVA nanofibers containing 17% SWNT could facilitate the colonization of brain-derived cells and U373 cell line in the scaffold and the efficient exchange of nutrients and metabolic wastes between the scaffold and its environment would happen [11, 13, 15].

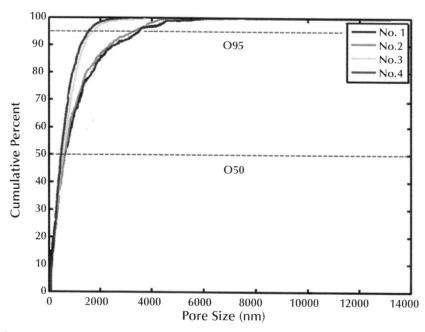

Figure 5.6. PSD curves of SWNT-CS/PVA nanofibrous membranes: No.1: Bulk CS/PVA, No. 2: CS/PVA with 7% SWNT, No. 3: CS/PVA with 12% SWNT, N0. 4: CS/PVA with 17% SWNT.

Mechanical Analysis

The versatility of electrospinning method gives room for scaffold improvements, like tuning the mechanical properties of the tubular structure or providing biomimetic functionalization. Figure 5.7 shows the tensile strength response recorded samples for CS/PVA and SWNT-CS/PVA nanofibers. As expected, it was observed that SWNT-CS/PVA blend presented higher mechanical properties in comparison to bulk CS/PVA nanofibers. It is important to point out that some processing parameters, for instance, the degree of acetylation of chitosan and blending procedure, will interact to influence the mechanical properties. In summary, as far as mechanical property is concerned, the results obtained for the SWNT-CS/PVA developed in this research is suitable for potential application as tissue replacement [31, 35, 36].

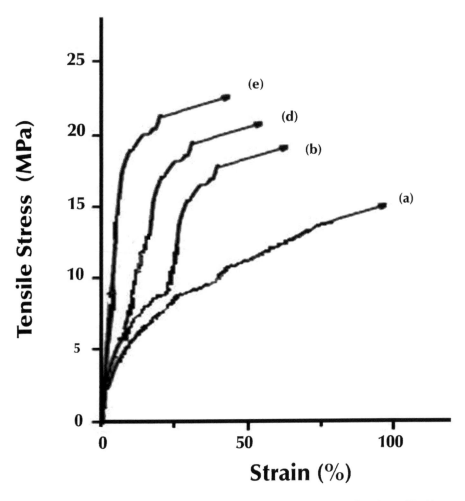

Figure 5.7. Tensile strength of SWNT-CS/PVA nanofibrous membranes: (a) Bulk CS/PVA, (b) CS/PVA with 7% SWNT, (c) CS/PVA with 12% SWNT, (d) CS/PVA with 17% SWNT.

Cell Viability Assessment

Cell viability was measured using MTT and NR assay. As shown in Fig. 5.8 and Fig. 5.9, SWNT-CS/PVA nanocomposites exhibited equal biocompatibility in comparison to control groups. In general, the proliferation rate of brain-derived cells and U373 cell line in nanocomposite extraction solutions was similar or even higher than negative (TPS) and positive (CS/PVA nanofibers) control groups. Although, the result showed that the both cell types have higher viability on SWNT-CS/PVA nanocomposites in the 7th day but there were no statistically significant differences (p<0.05) among SWNT-CS/PVA and control groups. Thus, the results suggested that the extracts obtained by SWNT-CS/PVA samples do not affect cell viability and their proliferation rates. Since both the topography and the porosity of a scaffold play a significant role in the proliferation of cells, it is assumed that the behavior of cells on SWNT-CS/PVA nano-composites controls the overall performance of this scaffold to be used as neural tissue engineering platform.

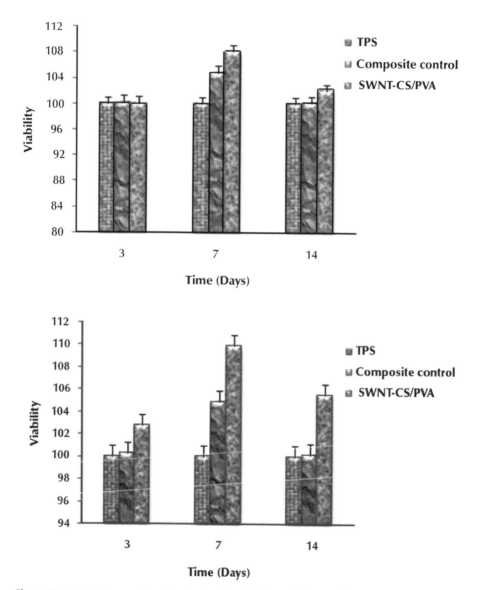

Figure 5.8. MTT assay results of the brain-derived cells and U373 cell lines on SWNT-CS/PVA nanocomposites at various incubation times. CS/PVA nanocomposites were considered as composite control and tissue culture polystyrene as negative control. (a) Cell proliferation of brain-derived cells. (b) Cell proliferation U373 cell lines. There were no statistically significant differences among the samples.

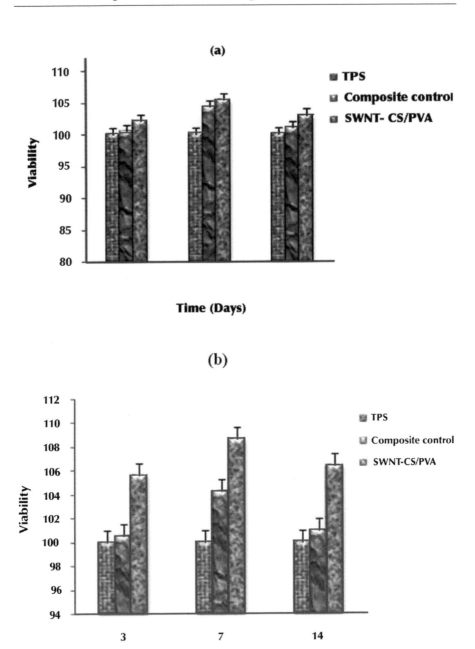

Figure 5.9. NR assay results of the brain-derived cells and U373 cell lines on SWNT-CS/PVA nanocomposites at various incubation times. CS/PVA nanocomposites were considered as composite control and tissue culture polystyrene as negative control. (a) Cell proliferation of brain-derived cells. (b) Cell proliferation U373 cell lines. There were no statistically significant differences among the samples.

SUMMARY

Chitosan/poly(vinyl alcohol) reinforced Single-walled carbon nanotube nanofibers were effectively prepared via electrospinning technique. This method is simple, fast and, above all, caused no damage or deformation to the polymer matrix as confirmed by Raman spectroscopy. Good adhesion between SWNTs and substrates blocks the loss of SWNTs when they are exposed outdoors, and leads to reliably strong performance of SWNTs applications. Here, the use of electrospinning to enhance the adhesion between SWNTs and polymer substrates is reported. SWNT-CS/PVA provides an optimal microenvironment with high tensile strength in terms of morphology, porosity, and molecular structure for the brain-derived cells and U373 cell lines to maintain relatively high biological activity and stability. Moreover, cell viability assays have given important evidence that the nanocomposite evaluated is non-toxic, and potentially biocompatible. In summary, these novel functional nanocomposites based on CS/PVA reinforced SWNT membranes have broadened the number of choices of biomaterials to be potentially used in neural tissue engineering.

ACKNOWLEDGMENT

The authors thank to the staffs of Cell Bank of Pasteur Institute of Iran for their cooperation and useful consultation. Financial support from National Cell Bank of Pasteur Institute of Iran and Guilan Science and Technology Park (GSTP) is also acknowledged.

KEYWORDS

- Chitosan
- Electrospun nanofibers
- Pore size distribution (PSD) curve
- Single-walled carbon nanotubes
- Ultrasonication

Chapter 6

Wireless Wearable ECG Monitoring System

V. Mottaghitalab and M. S. Motlagh

INTRODUCTION

According to world statistics, 29% of the global causes of death is heart disease. Once more, the importance of this issue and concern is based on published statistics. Statistics shows that before cardiac arrest, 22% of patients experience chest pain for 120 min, 15% suffered asthmatic for 30 min, 7% suffer nausea and vomiting for 120 min and 5% had been suffered from dizziness. Only 25% of patients had no warning signs before heart attack. The researchers reported that 90% of patients have some sort of signal at least 5 min before the symptoms. Typically, the electrocardiogram (ECG) is the first preliminary test on heart patients. This requires spending lots of time and cost to the patient. In addition, the received information from the patient's ECG shows the situation at test moment. Meanwhile, the heart had a different function in different physical conditions and everyday movement, which is highly encountered to risk of heart stroke. Thus continues access to the heart wave function during daily activities by receiving continuous bioelectric signals is essential.

Electrocardiogram (ECG) system is a sophisticated device for patient vital heart signal monitoring and supervision. An extensive range of human physiological conditions can be deduced from the PQRST waves obtained from an ECG instrument. A typical heart signal wave has been demonstrated in Fig. 6.1.

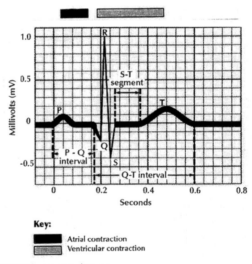

Figure 6.1. Typical PQRST heart signal wave.

P wave indicates electrical current in atriums, QRS complex shows electrical current in ventricles and T wave demonstrates heart relaxation period and charging time between two hear beats. The heart muscle tissue is the self-stimulating with spontaneous contractions that convert biochemical energy into kinetic energy, and it provide the needed driving force for the circulation system.

Heart muscle contractions associated with electrical stimulation is called the action potential and bioelectrical signal is triggered. Action potential causes electric shock that passes through cell membranes and controls the amount of contraction in heart muscle strings. Charged ions exchange and displacement from one side to the other side of the cell membrane that is carried by ion channels, causing action potential [1, 2, 3]. General release in the form of an action potential is shown in Fig. 6.2.

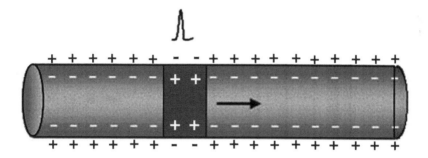

Figure 6.2. The release of action potential.

Membranes of heart cells control the number of ions including sodium, potassium, calcium and chloride ions into and out of the cell. The ions enter and exit take place through a special duct for the same ion. Ion channels transmission is capable for opening and closing due to the changing of cell voltage. In resting heart electrochemical equilibrium between the ions located on both sides of the cell membrane is established in which case the electric potential inside the cell than outside the cell is negative and its value is about 80 to 90 mV [4, 5]. By the time passing, the cell membrane conductivity changes through the opening and closing of the ions passing channels and consequently the potential difference between inside and outside the cell environment changes. In this process the cell initially is de-polarized and again with increasing of the polarity cell potential back to rest. This process is performed on all heart cells, resulting in the contraction of heart muscle and back again to initial form [6, 7].

Electrical activity of heart cells in the body produces electrical current on the skin resulting in the measurable potential difference. Consequently, the measuring of potential difference on skin surface is an effective way to evaluate the heart performance.

ECG machine combine the received bioelectrical signals from implemented sensors. The attached sensors to certain points of the body are able to detect and measure the changes in the potential difference created in the skin surface. The used sensors in the ECG should be highly electrically conductive because the variation of potential differences in skin surface is very low and is in order of milivolt (mV) [8]. In most cases, a combination of silver/silver chloride is used for fabrication of these sensors. Also for better contact with the skin, the conductive gel is used [9].

Acquired signals using sensors are quite weak and boosting system is extremely necessary to monitor the change in bioelectrical signals. These electrical signals also can be affected by electromagnetic waves, the change in user geographical position and connecting wires. It is therefore necessary to use filter and or noise depletion circuit. Two types of filters including high pass (HP) and low pass (LP) have been designed for commercial ECG. HP filters adjusts between 0.5 and 1 Hz and sort out those signals smaller than threshold limit and reduce the baseline deviation. LP filters however adjust on 40 Hz and remove signals higher than threshold limit. In addition, it can delete the environmental noises between 50 and 60 Hz [10]. The abovementioned items beside monitoring system make commercial ECG instruments are quite bulky and miniaturization in recent years has enabled development of wearable versions that collect and process ECG data. Subtle changes in the physiological signals of an individual can be monitored using wireless wearable ECG monitoring system. Telemetry in combination with such systems prepares physicians to take clinical decisions when considerable changes are noticed in the physiological parameters [11, 12]. The conventional physiological monitoring system used in hospitals cannot be used for wearable physiological monitoring applications due to the following reasons [13–14].

- The conventional physiological monitoring systems are massive to be used for wearable monitoring.
- The gels used in the electrodes dry out when used over a period of time, which lead to increase in the contact resistance and thereby degrading the signal quality.
- The gels used in the electrodes cause irritations and rashes when used for longer durations.
- There are number of hampering wires from the sensors to the data acquisition system.
- The sensors used in conventional monitoring systems are bulky and are not comfortable to wear for longer durations.

To overcome the above problems associated with the conventional physiological monitoring there is a need to develop sensors for wearable monitoring, integrate them into the fabric of wearer, and continuously monitor the physiological parameters

Looking at the recent trends in biomedical applications, a major advancement can be noted in non-stationary health monitoring devices. This ranges from simple and portable Holter instruments to sophisticated and costly implantable gadgets.

The Holter monitors have been used only to collect data in ambulatory patients. Processing and analysis are then performed offline on recorded data [14]. Systems with multiple sensors have too many wires between the sensors and the monitoring device, which limit the patient's activity and comfort level. Available systems also lack universal connectivity of interfacing to any output display device through common communication ports. There is a requirement of data acquisition (DAQ) circuits with analog to digital converters as the interface between the Instrument and the computer [15, 28].

The hospitals generally use Wilson Central Terminal arrangement having three electrodes placed at the limbs and connected at the inverting input of the ECG amplifier. The Wilson reference can however degrade the overall amplifier specifications [16]. Implantable Cardioverter Defibrillator (ICD) is other system, which is very expensive and invasive method to record physiological data. The ICD is used only on high risk cardiac patients [17]. MOLEC monitor is an embedded real time system that captures, processes, detects, analyzes and notifies abnormalities in ECG [18]. However, the cost of MOLEC monitor is high. Further, EPI-MEDIC is a twelve-lead ECG system that allows continuous monitoring, but has a large array of electrodes, which makes the system cumbersome [19]. Normal ECG done in clinical setting allows monitoring and recording but gives no analytical results. Available data analysis algorithms are complicated, as they do not implement transparent decision procedure.

The growing health concerns, especially for cardiac disorders reflect on the need of developing a simple and portable ECG system for daily life.

A number of wearable physiological monitoring systems have been put into practical use for health monitoring of the wearer in hospital and real life situations and their performances have been reported [20–22]. Varying degrees of success have been reported and the percentage failures in the outdoor use are high. The drawbacks associated with the conventional wearable physiological systems are

- The cables woven into the fabric to interconnect the sensors to the wearable data acquisition hardware pick up interfering noises (e.g., 50 Hz power line interference). The wires integrated into fabric act like antennas and can easily pick up the noises from nearby radiating sources.

- The sensors once integrated into the fabric, its location cannot be changed or altered easily.

- The power required for the sensors to operate is to be drawn from the common battery housed in the wearable data acquisition hardware and are routed through wires woven in the fabric.

- Typically these systems consist of a centralized processing unit to digitize, process and transmit the data to a remote monitoring system. The processor is loaded heavily to perform multi-channel data acquisition, processing and transmission of data.

- The cables from the vest interconnecting the sensors can get damaged very easily due to twisting and turning of the cables, while the wearer is performing his routine activity.

To overcome the above issues related to wearable physiological monitoring, the individual sensors integrated into the vest can be housed with electronics and wireless communication system to acquire and transmit the physiological data. The recent advances in the sensors (MEMS and Nanotechnology), low-power microelectronics and miniaturization and wireless networking enable Wireless Sensor Networks (WSN) for human health monitoring. A number of tiny wireless sensors, strategically placed on the human body create a wireless body area network that can monitor various vital signs, providing real-time feedback to the user and medical personnel [23–24].

The implications and potentials of these wearable health monitoring technologies are paramount, for their abilities to: (1) detect early signs of health deterioration, (2) notify health care providers in critical situations, (3) find correlations between lifestyle and health issues, (4) bring sports conditioning into a new dimension, by providing detailed information about physiological signals under various exercise conditions, and (5) transform health care by providing doctors with multi-sourced, real-time physiological data.

Regarding to these concerns textile material offers new possibilities to bring new advancement to ECG system using intelligent textiles. Previous work on so-called 'intelligent textiles' shows that biomedical monitoring systems can greatly benefit from the integration of electronics in textile materials [25, 26]. This synergy between textiles and electronics mainly originates from the fact that clothing is our most natural interface to the outside world. The higher the level of integration of the sensors and circuits in the clothing, the more unnoticeable the monitoring system becomes to its user and thus the higher the comfort of the user. The elements of wearable ECG system operationally is similar to commercial device but have some basic difference in ECG electrodes and the size of amplifying and filtering circuits.

Various technologies used to design and manufacturing of embedded sensors in smart textile based on their properties such as flexibility, durability, stability during use, and no damage to the skin. Using metallic yarns as a woven matrix or Textrod is the most common method of making textile sensors. In this method, stainless metallic yarns using weaving or knitting techniques can produce seamless texture [27].

Sensors made by steel yarn textures able to receive and transfer bioelectrical heart signals and transfer heart signals nonetheless these sensors are also have disadvantages. Woven sensors create some problems such as the abrasions on the skin surface from free tail of metallic yarn.

Moreover, the sewn sensors due to low flexibility provide not sufficient comfort for users. The problem can be resolved using circular seamless texture in the context of the fabric sensor, but the location for a specific user is considered by default. For example, a different user needs different cloth according to their size. In another attempt, the flexible sensor is prepared using graphite coated with nickel in the bed of silicon. The problem is connecting of the sensor to the fabric and its connection to the communications system using metallic wire for signal transmission. In accordance to sensor thickness, the user does not feel comfortable. Another way for making wearable sensor is sputter coating. The fabric sensor in this method is coated by copper particle. This sensor due to very low thickness shows good flexibility. Moreover, the fabricated sensor shows very good uniformity and an acceptable quality of received signals. The big problem is the high cost of producing and the low production rate.

In the development of smart textile sensors, another approach is presented which uses electroless metal reduction method for preparation of fabric sensors. This approach can be divided into two groups including electroless plating and electroless printing. The first technique achieves the textile sensor through making a highly conductive fabric. It in turn must be sewn on specified position to touch skin for gathering of bioelectric data. In the latter, however there is no need to sewing. In fact, the sensor as a pattern can be printed on fabric with a very high precision. The major benefit of using this technique is that the measurement devices can be 'seamlessly' integrated into textile. Hereby, both patient comfort and ease of use in everyday monitoring are enhanced. Also, a single channel ECG monitor has been developed that successfully extracts ECG wave in a simple, wireless and cost effective manner. The amplified and filtered output is then fed to a commercially available cost effective radio frequency short range Bluetooth transmitter integrated in a pocket PC equipped with an software for data analysis. Automatic alarm detection system, can also give early alarm signals even if the patient is unconscious or unaware of cardiac arrhythmias. Simultaneously, a warning message is sent to the doctor at the hospital uses a short message. Also GPRS enable us as to transmit ECG data in a specified time frame to a Clinical Diagnostic Station (CDS). This chapter describes the implementation of and experiences with this new system for wireless monitoring.

EXPERIMENTAL

Sensor Preparation Based on Electroless Plating

Cotton fabrics (53×48 count/cm^2, 140 g/m^2, taffeta fabric) with white color were used as substrates. The surface area of each specimen is 400 cm^2. Electroless plating was carried out by multi-step processes including scouring, rinsing, sensitization, rinsing, activation, rinsing, electroless silver plating, rinsing and drying. The fabric specimens (10 cm $\times$ 10 cm) were first scoured in non-ionic detergent (2 g/l) and $NaHCo_3$ (2 g/l) solution prior to use. The samples then were rinsed in distilled water. Next, surface sensitization was conducted by immersion of the samples into an aqueous solution containing $SnCl_2$ and HCl (40 ml/l 38% w/w). The specimens were again rinsed in distilled water and immersed in an activator containing $PdCl_2$ and HCl (20 ml/l, 38% w/w). The specimens were rinsed for third times in a large volume of deionized water for more than 5 min to prevent contamination of the plating bath. Then all samples immersed in the electroless bath containing silver nitrate, ammonium hydroxide, sodium hydroxide and α-glucose. The samples were rinsed in hot and cold water respectively, and then were dried in oven at 70°C.

Sensor Preparation Based on Electroless Printing

Screen-printing is a printing technique that uses a woven mesh to support an ink-blocking stencil. The attached stencil forms open areas of mesh that transfer ink or other printable materials, which can be pressed through the mesh as a sharp-edged image onto a substrate. A roller or squeegee is moved across the screen stencil, forcing or pumping ink past the threads of the woven mesh in the open areas. Screen-printing is also a stencil method of printmaking in which a design is imposed on a screen of silk or other fine mesh, with blank areas coated with an impermeable substance, and ink is forced through the mesh onto the printing surface. A typical white cotton T-shirt (53×48 count/cm^2, 140 g/m^2, taffeta fabric) as prototype marked on three small areas for further printing process of sensors. The surface area of each marked area is 25 cm^2. Electroless printing was carried out by multi-step processes like electroless plating except that the activation should be carried out on an specified sensor area. The whole t-shirt was first scoured in non-ionic detergent (2 g/l) and $NaHCo_3$ (2 g/l) solution prior to use. The sample then was rinsed in distilled water. Next, surface sensitization was conducted by immersion of the samples into an aqueous solution containing $SnCl_2$ and HCl (40 ml/l 38% w/w). The specimen then again rinsed in distilled water and prepared for printing process using a fine mesh screen. An activator containing $PdCl_2$ and HCl (20 ml/l, 38% w/w) is then prepared and its concentration increases by solvent vaporizing. During this step, viscosity continuously increases and heating process continues until viscosity reaches to 100 cP. Adjustment of viscosity is extremely necessary since the activator flows freely out of marked area for lower viscosity and also the screen mesh is blocked while the viscosity is higher than required level. The specimens were then rinsed for of the plating bath. Then activated area immersed in the electroless bath containing silver nitrate, ammonium hydroxide, sodium hydroxide and α-glucose. The printed area were rinsed in hot and cold water respectively, and then were dried in oven at 70°C.

General Plan for Wireless online Bio Monitoring and Alert System

Wearable physiological monitoring systems uses an array of sensors integrated into the fabric of the wearer to continuously acquire and transmit the physiological data to a remote monitoring station. The wearable monitoring systems allow an individual to monitor vital signs remotely and receive feedback to maintain a good health status. These systems alert medical personnel when abnormalities are detected. Generally, a wireless online bio monitoring and alert system composed of five main consoles. Figure 6.3 shows a general architecture for the bioelectrical signal acquisition and remotely sending to offshore medical database for further inspection. Simultaneously, this can be proposed for SMS bio alert system for emergency use.

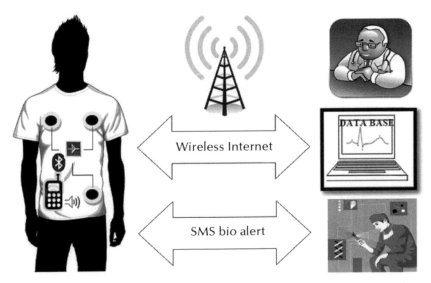

Figure 6.3. A typical image of wireless online bio monitoring and alert system.

The wearable physiological monitoring systems consist of five systems associated in a wireless data acquisition system.

- The vest with the screen printed and or integrated sensors
- The wearable data acquisition and processing hardware including amplifier filter circuit and Bluetooth unit for wireless data transfer
- Pocket /Pc receiver for additional filtering, amplifying, technical processing, and online transmission through GPRS protocol to CDS
- power generation and transmission unit for providing required energy to other units
- Remote offshore CDS as a database centre for data collection and precise medical investigation.

In the vest, sensors for acquiring the physiological parameters are integrated. The data acquired as the name implies, the three-electrode system uses only three electrodes to record the ECG. In such a system, the ECG is observed along one bipolar sensor between two of the electrodes while the third electrode serves as a ground. The sensors outputs and power line are interconnected to the data acquisition and processing hardware by means printed circuit using highly conducting pattern. In the wearable data acquisition and processing hardware, the circuits for amplification, filtering and digitization are housed. The digitized and processed data is transmitted wireless to a pocket PC. The ECG is continuously recorded with a built-in automatic alarm detection system, and the system can give early alarm signals even if the patient is unconscious or unaware of cardiac arrhythmias. Several alarm criteria where the doctor in a setup configuration can define the actual alarm limit; this includes bradycardia, tachycardia, and arrhythmia defined as variations in RR intervals. Our new concept has several advantages compare to existing solutions. If an abnormal ECG activity is encounter, the PDA will store one min of the ECG recording and then transmit the recording to the server. In addition, the PDA will calculate Heart Rate (HR) and variation in the R-R interval, averaged values together with maximum and minimum values are calculated every one minute.

In the remote offshore CDS the data is received and displayed in a form suitable for diagnosis. The doctor at the hospital uses a special remote client installed on standard PC .Trained personnel will thus be able to evaluate the ECG-recording, for diagnosing the conditions detected, and follow up the patient accordingly. The doctor can have a long-term ECG database of one patient. They are able to analyze the signal statically, and can detect an emergency condition. The required power is harvested from body thermal energy by means of textile-based thermopile based on Seeback effect. The produced energy during daily activity can be stored in fibrous super capacitor for usage in different element of wearable ECG monitoring system.

Physical Specifications of Fabric Sensors

The surface resistivity measurements of all samples using a homemade four-point probe test show that the resistance of cotton fabric has remarkable change after electroless plating. Figure 6.4 show the changes of fabric resistance after silver deposition in various concentrations of sensitization and activation agents. The obtained results demonstrated that the quality of silver plating has direct relation to concentration of sensitizer and activator. Furthermore, the results illustrated that the effective concentration of sensitizer was 10 g/l.

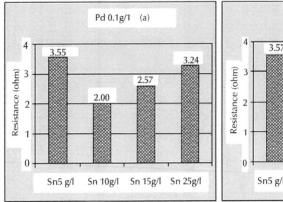
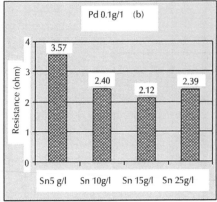

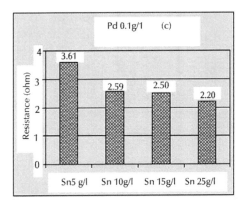

Figure 6.4. The Variation of resistance for Elctroless Fabric prepared in various concentration of sensitizer and (a) 0.1 g/l (b) 0.15 g/l (c) 0.2 g/l of activator.

It can also be stated that excess amount of activation agent compare to sensitization agent, reduce the conductivity of plated fabric due to contaminating of silver plating bath. A range of resistance between 2 and 4 Ω was achieved for electroless fabric which can be categorized as highly conductive material. Based on Fig. 6.4-a the lower resistance for conductive electroless fabric is achieved in an optimum concentration of 0.1 g/l of palladium as process activator. The sensor weight, thickness and stiffness before and after coating process were calculated and listed Table 6.1. The results show that coated fabric was heavier and thicker than original neat fabric. The measured percentage of changes in weight and thickness were 23 and 15 respectively. Thus it could be demonstrated that silver ions had stick to the fabric surface impressively

Table 6.1. Measurement of fabric weight, thickness and bending.

Specimen (10cm × 10cm)	Weight (g)	Thickness (mm)	Flexural rigidity (mg.cm)	
			Warp	Weft
Before plating	1.4	0.4	103.8	68.8
After plating	1.72(↑23%)	0.46(↑15%)	112(↑7.9%)	88.6(↑28.8%)

Figure 6.5 shows the uniform distribution of particle over sensor surface before and after electroless plating. A uniform layer of conductive material can be clearly seen without discontinuity wholly on filaments.

Figure 6.5. The SEM image of sensor substrate (a) before and (b) after electro plating.

The higher magnification picture has been captured to show the particle size and its distribution over sensor area (Fig. 6.6). A range of particle between 45 and 80 nm has been observed which indicate a nanoscale deposition of metallic particle. It is extremely important for bioelectric signal acquisition, which its intensity is a few mV. The more uniform conductive surface, the stronger electrical signal can be acquired.

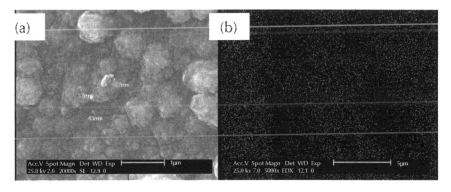

Figure 6.6. (a) SEM image of sensor surface at high magnification (b) WDX analysis for conductive nanoparticle distribution.

In addition, it has been demonstrated that the silver plated cotton fabric was stiffer than untreated one. Bending of fabric depend on surface friction between the fabric yarns and fibers. After electroless silver plating, presence of the silver particles on the fibers surface increase the friction between fibers and restrict their relative movement. Consequently, the stiffness of cotton fabric is increased after silver coating.

The estimation of fabric tearing strength, tearing elongation and crease are illustrated in Table 6.2. According to the obtained results the tearing strength and tearing elongation of cotton fabric decrease and increase respectively after silver plating. The decreasing of tearing strength and increasing of tearing elongation are the consequence of fabric exposure to highly low pH medium during sensitization and activation process. The cotton fiber inter bonding will be damaged in low pH value resulting the weakness of the cotton fabric. The values of crease recovery angle shows a significant improvement for crease resistance property of cotton fabric after silver electroless plating in both warp and weft direction.

Table 6.2. Estimation of fabric tearing strength, tearing elongation and crease.

	Tearing strength (kg.f)	Tearing elongation (mm)	Crease Recovery (degree°)	
			Warp	Weft
Before plating	1667.1	5.47	35	30
After plating	1553.7(↓6.8%)	8.05(↑47.2%)	40(↑14.3%)	32(↑6.67%)

Although the friction between fabric yarns and fibers increased after plating and it seems that the crease recovery angle should be decrease but in cotton fabric presence of silver particle on the surface of fibers prevent to creation of the hydrogen bond between cotton fibers during wrinkle implementation. Totally according to obtained results silver plating on cotton fabric decrease retained crease and enhance wrinkle recovery of cotton fabric.

Table 6.3 summarized the results of estimation of color change after washing, rubbing and perspiration. According to the obtained results washing and perspiration have no influence on the color of plated fabric. The color fastness results of washing and perspiration are rated as grade 5. The result of rubbing fastness reveals that color of coated fabric partially changed after dry rubbing. This approved that some silver particles were detached during dry rubbing; likewise, the results show wet rubbing fastness of plated fabric was comparatively poor considering commercial necessity.

Table 6.3. Estimation of color change.

Washing	Rubbing		Perspiration	
	Dry	Wet	pH 5	pH 8
5	3–4	3	5	5

In this work catalytic sites on the fabric surface were created by proceeding activation process in which palladium ions are reduced on the surface by stannous ions to a palladium layer that act as a catalyst stead stannous layer for the ensuing electroless deposition of silver. Therefore, in comparison of results it could be illustrated significant development in the surface resistance of electroless silver plating on the cotton fabric by using palladium chloride as an activator agent stead stannous chloride. The effect of washing on the surface resistance of plated fabric is illustrated in Fig. 6.7. As shown in Fig. 6.1, although the surface resistance of plated fabric after each washing repetition became less, the slope of resistance increasing is downward; furthermore, the change in surface resistance of coated fabric was negligible.

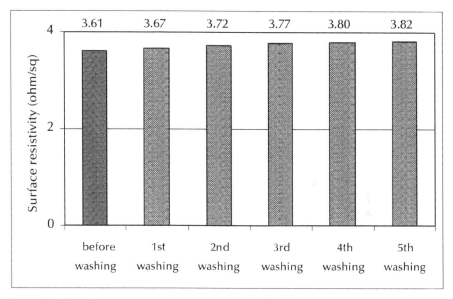

Figure 6.7. Effect of washing on the surface resistance of silver plated cotton fabric.

Figure 6.8 shows the influence of both dry and wet rubbing on the surface resistance of silver plated cotton fabric. According to the results shown in Fig. 6.8; by reiteration of both dry and wet rubbing test the difference of surface resistance between first and fifth test is insignificant, but meaningful difference appeared in surface resistance of silver plated fabric observed after first dry and wet rubbing test compared to untested one.

The surface resistance of coated fabric before rubbing test was 3.61 Ω/sq which decreased after dry and wet rubbing test to 4.74 Ω/sq and 6.10 Ω/sq respectively which indicate detaching of silver particles from fabric surface in first rubbing test; nevertheless, the silver plated cotton fabric has considerable surface resistance than uncoated one.

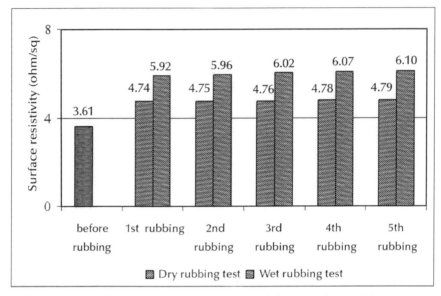

Figure 6.8. Influence of rubbing on the surface resistance of silver coated cotton fabric.

In the case of perspiration affect on the surface resistance of plated fabric, according to Fig. 6.9, the obtained results confirm that perspiration influence on surface resistance in both acidic and basic condition was inconsiderable as the surface resistance difference of silver plated cotton fabric after both acidic and basic perspiration has no change than untested one.

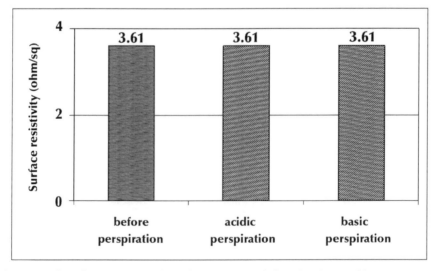

Figure 6.9. Effect of perspiration on the surface resistance of silver plated cotton fabric.

The Printed Circuit

Proposed plan also offers a new technique for signal transfer without using metallic wire and therefore it is free from challenges regarding to its connection to electronic hardware through soldering and or conductive glue. Printing mechanism is extremely precise to produce a sharp and quite straight line between two points, which sometimes are not close. Highly conductive silver ink have been prepared and printed on patterned area.

Internal Processor

Acquired data by embedded sensor arrays must be sent to an internal processor. In this system, data gathered by the sensors can be processed within the network and only aggregated information is sent to Pocket PC. In many applications, internal network processing can reduce the amount of data sent to the onshore terminal significantly and plays a major role in reducing communication power consumption. Moreover, in multi-hop networks, data transmission to a base station/terminal increases the power consumption of intermediate nodes and can inject delay to the data processing procedures. Many of these issues can be addressed using in-network processing. Fig. 6.10 shows the truly small internal processor chip equipped with commercially available cost effective radio frequency short range Bluetooth transmitter.

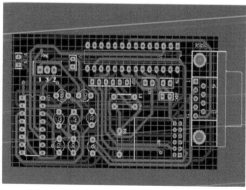

Figure 6.10. Internal processor including filter and amplifier.

The amplifier circuit has two outputs including about five times and 10 times amplified outputs. The amplified signals after removing disorders sends for Storage and further analysis to pocket PC. The on-body terminal collects the sensed data from wireless senor array and acts as a temporary storage unit. A potential benefit of the on-body terminal is the ability for the data to be remotely monitored by a healthcare provider without specific action by the patient. The acquired signals without any filtration in a short time frame has been demonstrated in Fig. 6.11. It is clearly indicated that the PQ and ST segments of heart signals cannot be distinguished and it has been dominated by noises originated from environment and or body movements.

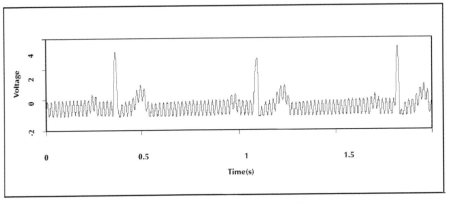

Figure 6.11. The 10 times acquired signals by pocket PC without filtration.

However, the observed quality for the PQRST wave after filtration and amplifying is much better and only some noises appears in PQ segments which is originates from environments.

Displaying of heart signals on pocket PC (Figs. 6.11 and 6.12) prove a good connection between sensor and internal processor and in turn confirm the data transfer between wireless sensor array and onshore terminal.

The final leg in this design hierarchy is the central diagnostic server. The server collects the data received from multiple on-body terminals and functions as a central storage unit. Moreover, extensive data analysis and processing are scheduled to be performed at the central server. The data can be "automatically" uploaded to the CDS for review. Therefore, the stored/pre-processed data can be constantly uploaded to a destination through possible underlying wireless infrastructures. Three modes of wireless communication, Bluetooth, wireless local area networks (WLAN) and general packet radio service (GPRS), can be used to enable such connections via the On-Body terminal device. GPRS allows data rates of 115 kbps (0.115 Mb/s) and, theoretically, of up to 160 kbps (0.16 Mb/s) on the physical layer. GPRS is capable of offering data rates of 384 kbps (0.384 Mb/s) and, theoretically, of up to 473.6 kbps (0.4736 Mb/s). In comparison, WLAN has a superior data rate (e.g., 54 Mb/s for 802.11 g) and a cheaper connection cost, but smaller coverage area and limited availability.

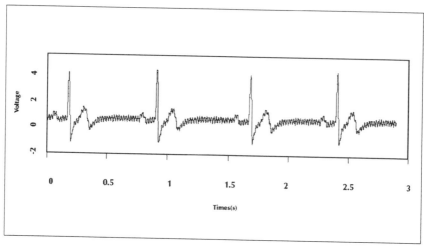

Figure 6.12. The 10 times acquired signals by pocket PC without filtration.

Figure 6.13 demonstrates real time online bioelectrical signals acquired by developed system in NRB CO (LTD) based on local GPRS facility in different rooms of a building.

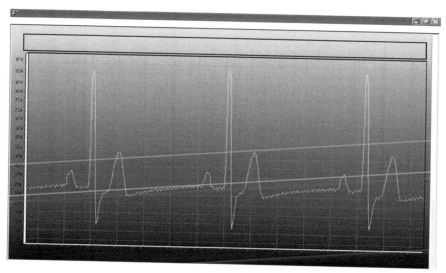

Figure 6.13. Typical ECG-recording from receiver unit connected to server.

Our system acts as a continuous event recorder, which can be used to follow up people who have survived cardiac arrest, ventricular tachycardia or cardiac syncope-in both ambulatory settings and in hospitals. It therefore seems reasonable to assume that our new ECG-monitoring system will be able to, reliably, detect rarely occurrences of cardiac arrhythmias. Thus make correct diagnosis even under situations where the patient has the ability to carry out daily life.

SUMMARY

A wearable wireless platform has been developed to collect the ECG signals and sending remotely in a real time frame. The fabrication process for sensor arrays uses electroless metal reduction. Two approaches including electroless plating and electroless printing have been employed. A highly conductive fabric for sewing on specified position employed for first approach. However the sensor as a pattern can be printed on fabric with a very high precision in second method. The major benefit of using this technique is that the measurement devices can be 'seamlessly 'integrated into textile for both patient comfort and ease of use in everyday life. In addition, a single channel wireless ECG system has been developed for successful data acquisition and remote sending. The amplified and filtered output is then fed to a commercially available cost effective radio frequency short range Bluetooth transmitter integrated in a pocket PC equipped with a software for data analysis. A fully automatic detection system can also give early alarm signals even if the patient is unconscious or unaware of cardiac arrhythmias. Simultaneously, a warning message is sent to the doctor as a short message. Also GPRS enable us as to transmit ECG data in a specified time frame to a Clinical Diagnostic Station (CDS).

KEYWORDS

- **Electrocardiogram**
- **Electroless printing**
- **MOLEC monitor**
- **Screen-printing**
- **Wearable monitoring systems**

Chapter 7

Conductive Chitosan Nanofiber

Z. M. Mahdieh, V. Mottaghitalab, N. Piri, and A K. Haghi

INTRODUCTION

Over the recent decades, scientists interested to creation of polymer nanofibers due to their potential in many engineering and medical properties [1]. According to various outstanding properties such as very small fiber diameters, large surface area per mass ratio, high porosity along with small pore sizes and flexibility, electrospun nanofiber mats have found numerous applications in diverse areas. For example in biomedical field nanofibers plays a substantional role in tissue engineering [2], drug delivery [3], and wound dressing [4]. Electrospinning is a novel and efficient method by which fibers with diameters in nanometer scale entitled as nanofibers, can be achieved. In electrospinning process, a strong electric field is applied on a droplet of polymer solution (or melt) held by its surface tension at the tip of a syringe needle (or a capillary tube). As a result, the pendent drop will become highly electrified and the induced charges are distributed over its surface. Increasing the intensity of electric field, the surface of the liquid drop will be distorted to a conical shape known as the Taylor cone [5]. Once the electric field strength exceeds a threshold value, the repulsive electric force dominates the surface tension of the liquid and a stable jet emerges from the cone tip. The charged jet is then accelerated toward the target and rapidly thins and dries as a result of elongation and solvent evaporation. As the jet diameter decreases, the surface charge density increases and the resulting high repulsive forces split the jet to smaller jets. This phenomenon may take place several times leading to many small jets. Ultimately, solidification is carried out and fibers are deposited on the surface of the collector as a randomly oriented nonwoven mat [6–7]. Figure 7.1 shows a schematic illustration of electrospinning setup.

The physical characteristics of electrospun nanofibers such as fiber diameter depend on various parameters which are mainly divided into three categories: solution properties (solution viscosity, solution concentration, polymer molecular weight, and surface tension), processing conditions (applied voltage, volume flow rate, spinning distance, and needle diameter), and ambient conditions (temperature, humidity, and atmosphere pressure) [9]. Numerous applications require nanofibers with desired properties suggesting the importance of the process control. This end may not be achieved unless having a comprehensive outlook of the process and quantitative study of the effects of governing parameters. In this context, Sukigara et al. [10] were assessed the effect of concentration on diameter of electrospun nanofibers. They indicated that the silk nanofibers diameter increases with increasing concentration.

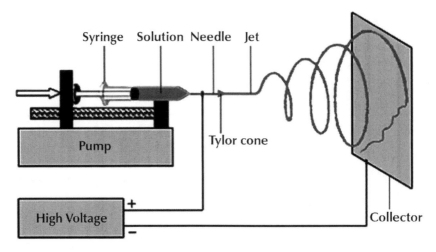

Figure 7.1. A typical image of electrospinning process [8].

Beside physical characteristics, medical scientists showed a remarkable attention to biocompatiblity and biodegredablity of nanofibers made of biopolymers such as collagen [11], fibrogen [12], gelatin [13], silk [14], chitin [15] and chitosan [16]. Chitin is the second abundant natural polymer in the world and Chitosan (poly-(1-4)-2-amino-2-deoxy-β-D-glucose) is the deacetylated product of chitin [17]. CHT is well known for its biocompatible and biodegradable properties [18].

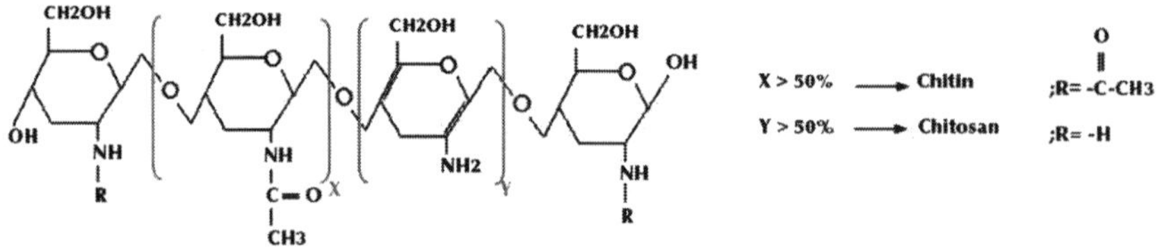

Scheme 7.1. Chemical structures of chitin and chitosan biopolymers.

Chitosan is insoluble in water, alkali, and most mineral acidic systems. However, though its solubility in inorganic acids is quite limited, chitosan is in fact soluble in organic acids, such as dilute aqueous acetic, formic, and lactic acids. Chitosan also has free amino groups which make it a positively-charged polyelectrolyte. This property makes chitosan solutions highly viscous and complicates its electrospinning [19]. Furthermore, the formation of strong hydrogen bonds in a 3-D network prevents the movement of polymeric chains exposed to the electrical field [20].

Different strategies have been used for bringing chitosan in nanofiber form. The three top most abundant techniques includes blending of favorite polymers for electrospinning process with CHT matrix [21–22], alkali treatment of CHT backbone to improve electro spinnability through reducing viscosity [23] and employment of concentrated organic acid solution to produce nanofibers by decreasing of surface tension [24]. Electrospinning of Polyethylene oxide (PEO)/CHT [21] and polyvinyl alcohol (PVA)/CHT [22] blended nanofiber are two recent studies based on first strategy. In second protocol, the molecular weight of chitosan decreases through alkali treatment. Solutions of the treated chitosan in aqueous 70–90% acetic acid produce nanofibers with appropriate quality and processing stability [23].

Using concentrated organic acids such as acetic acid [24] and triflouroacetic acid (TFA) with and without dichloromethane (DCM) [25–26] reported exclusively for producing neat CHT nanofibers. They similarly reported the decreasing of surface tension and at the same time enhancement of charge density of chitosan solution without significant effect on viscosity. This new method suggested significant influence of the concentrated acid solution on the reducing of the applied field required for electrospinning.

The mechanical and electrical properties of neat CHT electrospun natural nanofiber mat can be improved by addition of the synthetic materials including carbon nanotubes (CNTs) [27]. CNTs are one of the important synthetic polymers that were discovered by Iijima in 1991 [28]. CNTs either single walled nanotubes (SWNTs) or multi-walled nanotubes (MWNTs) combine together the physical properties of diamond and graphite. They are extremely thermally conductive like diamond and appreciably electrically conductive like graphite. Moreover, the flexibility and exceptional specific surface area to mass ratio can be considered as significant properties of CNTs [29]. The scientists are becoming more interested to CNTs for existence of exclusive properties such as superb conductivity [30] and mechanical strength for various applications. To our knowledge, there has been no report on electrospinning of CHT/MWNTs blend, except for several reports [30–31] that use PVA to improve spinnability. CNTs grafted by CHT were fabricated by electrospinning process. In these novel sheath-core nanofibers, PVA aqueous solution has been used for enhancing nanofiber formation of MWNTs/CHT. Results showed uniform and porous morphology of the electrospun membrane. Despite adequate spinnability, total removing of PVA from nanofiber structure to form conductive substrate is not feasible. Moreover, the structural morphology and mechanical stiffness is extremely affected by thermal or alkali solution treatment of CHT/PVA/MWNTs nanofibers. The chitosan/carbon nanotube composite can be produced by the hydrogen bonds due to hydrophilic positively charged polycation of chitosan due to amino groups and hydrophobic negatively charged of carbon nanotube due to carboxyl, and hydroxyl groups [32–34].

In current study, it has been attempted to produce a CHT/MWNTs nanofiber without association of any type of easy electro spinnable polymers. Also a new approach has been explored to provide highly stable and homogenous composite spinning solution of CHT/MWNTs in concentrated organic acids. This in turn present a homogenous conductive CHT scaffolds which is extremely important for biomedical implants.

EXPERIMENTAL

Materials

Chitosan polymer with degree of deacetylation of 85% and molecular weight of 5×10^5 was supplied by Sigma-Aldrich. The MWNTs, supplied by Nutrino, have an average diameter of 4 nm and purity of about 98%. All of the other solvents and chemicals were commercially available and used as received without further purification.

Preparation of CHT-MWNTs Dispersions

A Branson Sonifier 250 operated at 30 W was used to prepare the MWNT dispersions in CHT/organic acid (90% wt acetic acid, 70/30 TFA/DCM) solution based on different protocols. In first approach (current work) for preparation of sample 1, same amount (3 mg) as received MWNTs were dispersed into deionized water or DCM using solution sonicating for 10 min. Different amount of CHT was then added to MWNTs dispersion for preparation of a 8–12 wt% solution and then sonicated for another 5 min. Figure 7.2 shows two different protocols used in this study.

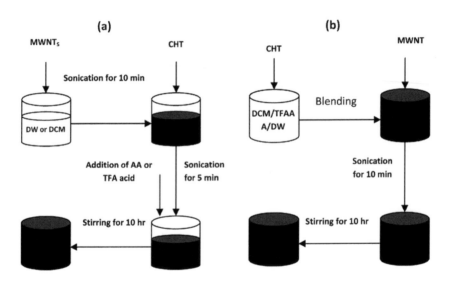

Figure 7.2. Two protocols used in this study for preparation of MWNTs/CHT dispersion (a) Current study (b) Ref. [35].

In next step, organic acid solution was added to obtain a CHT/MWNT solution with total volume of 5 mL and finally the dispersion was stirred for another 10 hr. The sample 2 was prepared using second technique. Same amount of MWNTs were dispersed in chitosan solution, and the blend with total volume of 5 mL were sonicated for 10 min and dispersion was stirred for 10 hr [35].

Electrospinning of Chitosan/Carbon Nanotube Dispersion

After the preparation of spinning solution, it was transferred to a 5 mL syringe and became ready for spinning of nanofibers. The experiments were carried out on a horizontal electrospinning setup shown schematically in Fig. 7.1. The syringe containing CHT/MWNTS solution was placed on a syringe pump (New Era NE-100) used to dispense the solution at a controlled rate. A high voltage DC power supply (Gamma High

Voltage ES-30) was used to generate the electric field needed for electrospinning. The positive electrode of the high voltage supply was attached to the syringe needle via an alligator clip and the grounding electrode was connected to a flat collector wrapped with aluminum foil where electrospun nanofibers were accumulated to form a nonwoven mat. The voltage and the tip-to-collector distance were fixed respectively on 18–24 kV and 4–10 cm. The electrospinning was carried out at room temperature. Subsequently, the aluminum foil was removed from the collector.

Measurements and Characterizations

A small piece of mat was placed on the sample holder and gold sputter-coated (Bal-Tec). Thereafter, the micrograph of electrospun PVA fibers was obtained using scanning electron microscope (SEM, Phillips XL-30). Fourier transform infrared spectra (FTIR) were recorded using a Nicolet 560 spectrometer to investigate the interaction between CHT and MWNT in the range of 800–4000 cm^{-1} under a transmission mode. The size distribution of the dispersed particle was evaluated with a Zetasizer (Malvern Instruments). The conductivity of the conductive fibers was measured using the four point-probe technique. A homemade four probe electrical conductivity cell operated at constant humidity has been employed. The electrodes were circular pins with separation distance of 0.33 cm and fibers were connected to pins by silver paint (SPI). Between the two outer electrodes a constant DC current was applied by Potentiostat/Galvanostat model 363 (Princeton Applied Research). The generated potential difference between the inner electrodes along the current flow direction was recorded by digital multimeter 34401A (Agilent). The conductivity (δ: S/cm) of the nanofiber thin film with rectangular surface can then be calculated according to length (L:cm), width (W:cm),thickness (t:cm), DC current applied (mA)and the potential drop across the two inner electrodes (mV). All measuring repeated at least five times for each set of samples.

$$\delta = \frac{I \times L}{V \times W \times t} \tag{1}$$

RESULTS AND DISCUSSION

The Characteristics of MWNT/CHT Dispersion

Utilisation of MWNTs in biopolymer matrix initially requires their homogenous dispersion in a solvent or polymer matrix. Dynamic light scattering (DLS) is a sophisticated technique used for evaluation of particle size distribution. DLS provides many advantages as a particle size analyzes method that measures a large population of particles in a very short time period, with no manipulation of the surrounding medium. Dynamic light scattering of MWNTs dispersions indicate that the hydrodynamic diameter of the nanotube bundles is between 150 and 400 nm after 10 min of sonication for sample 2. (Fig. 7.4)

MWNTs bundle in sample 1 (different approach but same sonication time compared to sample 2) shows a range of hydrodynamic diameter between 20 and 100 nm. (Fig. 7.4).The lower range of hydrodynamic diameter for sample 1 can be correlated to more exfoliated and highly stable nanotubes strands in CHT solution. The higher stability of sample 1 compared to sample 2 over a long period of time is confirmed by solution stability test. The results presented in Fig. 7.3 indicate that procedure employed for preparation of sample 1 (current work) was an effective method for dispersing MWNTs in CHT/acetic acid solution. However, MWNTs bundles in sample 2 was found to re-agglomerate upon standing after sonication, as shown in Fig. 7.5 where indicate the sedimentation of large agglomerated particles.

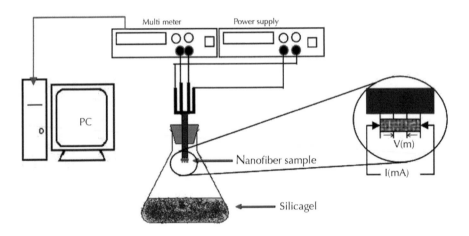

Figure 7.3. The experimental setup for four probe electrical conductivity measurement of nanofiber thin film.

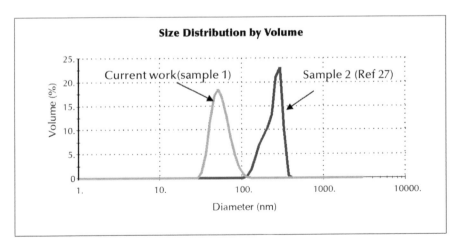

Figure 7.4. Hydrodynamic diameter distribution of MWNT bundles in CHT/acetic acid (1%) solution for different preparation technique.

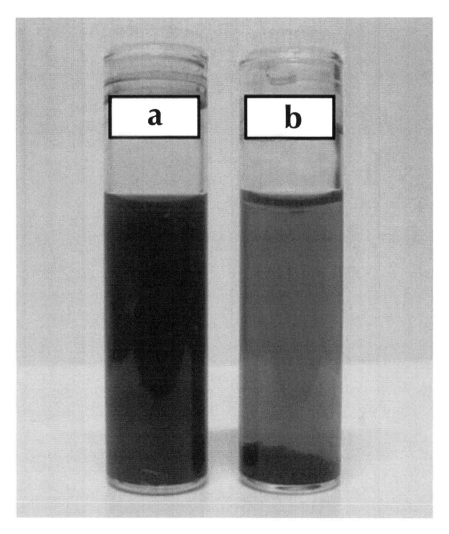

Figure 7.5. Stability of CHT-MWNT dispersions (a) Current work (sample 1) (b) Ref [35] (sample 2).

Despite the method reported in Ref. 27 neither sedimentation nor aggregation of the MWNTs bundles was observed in first sample. Presumably, this behavior in sample 1 can be attributed to contribution of CHT biopolymer to forms an effective barrier against reagglomeration of MWNTs nano particles. In fact, using sonication energy, in first step without presence of solvent, make very tiny exfoliated but unstable particle in water as dispersant. Instantaneous addition of acetic acid as solvent to prepared dispersion and long mixing most likely helps the wrapping of MWNTs strands with CHT polymer chain.

Figure 7.6 shows the FTIR spectra of neat CHT solution and CHT/MWNTs dispersions prepared using strategies explained in experimental part. The interaction between MWNTs and CHT in dispersed form has been understood through recognition of functional groups. The enhanced peaks at ~1600 cm⁻¹ can be attributed to (N-H) band and (C=O) band of amid functional group. However the intensity of amid group for CHT/MWNTs dispersion has been increased presumably due to contribution of G band in MWNTs. More interestingly, in this region, the FTIR spectra of MWNTs-CHT dispersion (sample 1) have been highly intensified compared to sample 2 [35]. It can be correlated to higher chemical interaction between acid functionalized C-C group of MWNTs and amid functional group in CHT.

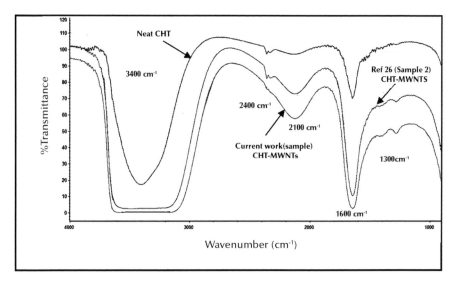

Figure 7.6. FTIR spectra of CHT-MWCNT in 1% acetic acid with different techniques of dispersion.

This probably is the main reason of the higher stability and lower MWNTs dimension demonstrated in Figs. 7.4 and 7.5. Moreover, the intensity of protonated secondary amine absorbance at 2400 cm⁻¹ for sample prepared by new technique is negligible compared to sample 2 and neat CHT. Furthermore, the peak at 2123 cm⁻¹ is a characteristic band of the primary amine salt, which is associated to the interaction between positively charged hydrogen of acetic acid and amino residues of CHT. Also, The broad peaks at ~3410 cm⁻¹ due to the stretching vibration of OH group superimposed on NH stretching bond and broaden due to inter hydrogen bonds of polysaccharides. The broadest peak of hydrogen bonds was observed at 3137–3588 cm^{-1} for MWNTs/CHT dispersion prepared by new technique (sample 1).

The Physical and Morphological Characteristics of MWNTs/CHT Nanofiber

The different solvents including acetic acid 1–90%, pure formic acid, and TFA/DCM tested for the electrospinning of chitosan/carbon nanotube. No jet was seen upon applying the high voltage even above 25 kV by using of acetic acid 1–30% and formic acid as the solvent for chitosan/carbon nanotube. When the acetic acid 30–90%, used as the solvent, beads were deposited on the collector. Therefore, under these conditions, an electrospun fiber of carbon nanotube/chitosan could not be obtained (data not shown).

Figure 7.7 shows scanning electronic micrographs of the MWNTs/CHT electrospun nanofibers in different concentration of CHT in TFA/DCM (70:30) solvent. As presented in Fig. 7.7a, at low concentrations of CHT the beads deposited on the collector and thin fibers coexited among the beads. When the concentration of CHT was increased as shown in Fig. 7.7a–c the beads was decreased. Figure 7.7c show homogenous electrospun nanofibers with minimum beads, thin fibers and interconnected fibers. More increasing of concentration of CHT lead to increasing of interconnected fibers at Fig. 7.7d–e. Figure 7.8 show the effect of concentration on average diameter of MWNTs/CHT electrospun nanofibers. Our assessments indicate that the fiber diameter of MWNTs/CHT increases with the increasing concentration. In this context, there are several studies that have reported similar to our results [36–37]. Hence, MWNTs/CHT solution in TFA/DCM (70:30) with 10 wt% chitosan resulted as optimization conditions of concentration for electrospinning. An average diameter of 275 nm (Fig. 7.7c: diameter distribution, 148–385) investigated for this conditions. Table 7.1 list the variation of nanofiber diameter and four probe electrical conductivity based on the different loading of CHT. One can expect the lower conductivity one the CHT content increases. However, this effect has been damped by decreasing of nanofiber diameter. This led to a nearly constant conductivity over entire measurements.

To understanding the effects of voltage on morphologies of CHT/MWNT electrospun nanofibers, the SEM images at Fig. 7.9 were analyzed. In our experiments, 18 kv were attained as threshold voltage, where fiber formation occurred. When the voltage was low the beads and some little fiber deposited on collector (Fig. 7.9a). As shown in Fig. 7.9a–d, the beads decreased by increasing voltage from 18 kV to 24 kV for electrospinning of fibers. The nanofibers collected by applying 18 kV (7.9a) and 20 kV (7.9b) were not quite clear and uniform. The higher the applied voltage, the more uniform nanofibers with less distribution starts to form. The average diameter of fibers, 22 kV (7.9c), and 24 kV (7.9d), respectively, were 204 (79–391), and 275 (148–385).

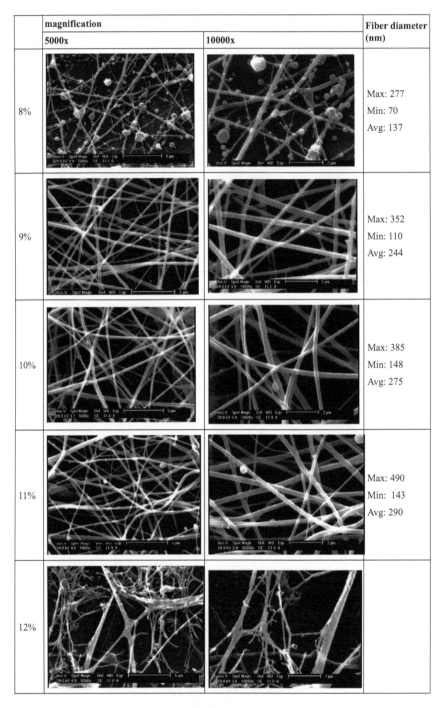

magnification		Fiber diameter (nm)
5000x	**10000x**	
8%		Max: 277 Min: 70 Avg: 137
9%		Max: 352 Min: 110 Avg: 244
10%		Max: 385 Min: 148 Avg: 275
11%		Max: 490 Min: 143 Avg: 290
12%		

Figure 7.7. Scanning electron micrographs of electrospun nanofibers at different CHT concentration (wt%): (a) 8, (b) 9, (c) 10, (d) 11, (e) 12, 24 kV, 5 cm, TFA/DCM: 70/30, (0.06% wt MWNTs).

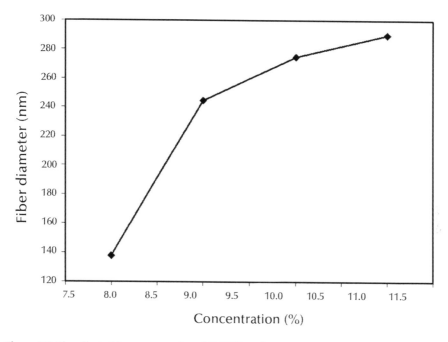

Figure 7.8. The effect of the concentration of CHIT/CNT dispersion on fiber diameter.

Table 7.1. The variation of conductivity and mean nanofiber diameter versus Chitosan loading.

% CHT (%w/v)	% MWNT (%w/v)	Voltage (KV)	Tip to collector (cm)	Diameter (nm)	Conductivity (S/cm)
8	0.06	24	5	137 ± 58	NA
9	0.06	24	5	244 ± 61	9×10^{-5}
10	0.06	24	5	275 ± 70	9×10^{-5}
11	0.06	24	5	290 ± 87	8×10^{-5}
12	0.06	24	5	Non uniform	NA

The conductivity measurement given in Tables 7.2, 7.3 and 7.4 confirms our observation in first set of conductivity data. As can be seen from last row, the amount of electrical conductivity reaches to a maximum level of 9×10^{-5} at prescribed setup.

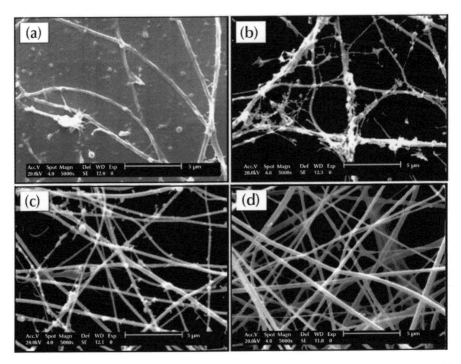

Figure 7.9. Scanning electronic micrograghs of electrospun fibers at different voltage (kV): (a) 18, (b) 20, (c) 22, (d) 24, 5 cm, 10 wt%, TFA/DCM: 70/30. (0.06% wt MWNTs).

Table 7.2. The variation of conductivity and mean nanofiber diameter versus applied voltage.

% CHT (%w/v)	% MWNT (%w/v)	Voltage (KV)	Tip to collector (cm)	Diameter (nm)	Conductivity (S/cm)
10	0.06	18	5	Non uniform	NA
10	0.06	20	5	Non uniform	NA
10	0.06	22	5	201 ± 66	6×10^{-5}
10	0.06	24	5	275 ± 70	9×10^{-5}

Table 7.3. The variation of conductivity and mean nanofiber diameter versus applied voltage.

% CHT (%w/v)	% MWNT (%w/v)	Voltage (KV)	Tip to collector (cm)	Diameter (nm)	Conductivity (S/cm)
10	0.06	24	4	Non uniform	NA
10	0.06	24	5	275 ± 70	9×10^{-5}
10	0.06	24	6	170 ± 58	6×10^{-5}
10	0.06	24	7	132 ± 53	7×10^{-5}
10	0.06	24	8	Non uniform	NA
10	0.06	24	10	Non uniform	NA

The distance between the tip to collector is another approach that controls the fiber diameter and morphology. Figure 7.10 shows the change in morphologies of CHT/MWNTs electrospun nanofibers at different distance between the tip to collector. When the distance tip-to-collector is not long enough, the solvent could not be vaporized, hence, a little interconnected thick fiber deposits on the collector (Fig. 7.10a). In the 5 cm distance of tip-to-collector (Fig. 7.10b) rather homogenous nanofibers have obtained with negligible beads and interconnected fibers. However, the beads increased by increasing of distance tip-to-collector as represented from Fig. 7.10b to 7.10f. Similar results was observed for chitosan nanofibers fabricated by Geng et al. [24]. Also, the results show that the diameter of electrospun fibers decreased by increasing of distance tip to collector in Fig. 7.10b, 7.10c, 7.10d, respectively, 275 (148–385), 170 (98–283), 132 (71–224). Similar effect of distance between tip to collector on fiber diameter has observed in previous studies [38–39]. A remarkable defects and nonhomogenity appears for those fibers prepared at a distance of 8 cm (Fig. 7.10e) and 10 cm (Fig. 7.10f). However, 5 cm for distance tip-to-collector was seen proper for electrospinning.

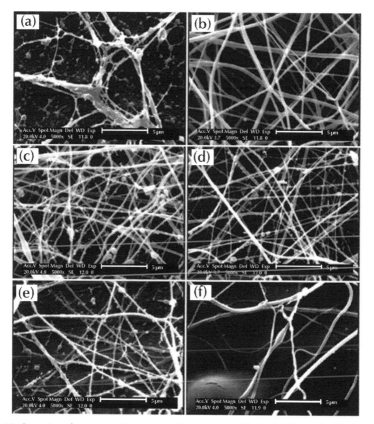

Figure 7.10. Scanning electronic micrograghs of electrospun fibers of Chitosan/Carbon nanotubes at different tip-to-collector distances (cm): (a) 4, (b) 5, (c) 6, (d) 7, (e) 8, (f) 10, 24 kV, 10 wt%, TFA/DCM: 70/30.

Conductivity results also are in agreement with those data obtained in previous parts. The nonhomogenity and huge bead densities plays as a barrier against electrical current and still a bead free and thin nanofiber mat shows higher conductivity compared to other samples. Experimental framework in this study was based on parameter adjusting for electrospinning of conductive CHT/MWNTs nanofiber. It can be expected that, the addition of nanotubes can boost conductivity and also change morphological aspects which is extremely important for biomedical applications.

SUMMARY

Conductive composite nanofiber of CHT/MWNTs has been produced using conventional electrospinning technique. A new protocol suggested for prepararation of electrospinning solution which shows much better stability and homogeneity compared previous techniques. Several solvent including acetic acid 1–90%, formic acid, and TFA/DCM (70:30) were investigated in the electrospinning of CHT/MWNTs dispersion. It was observed that the TFA/DCM (70:30) solvent is most preferred for fiber formation process with acceptable electrospinnability. The formation of nanofibers with conductive pathways regarding to exfoliated and interconntected nanotube strands is a breakthrough in chitosan nanocomposite area. This can be considered as a significant improvement in electrospinning of chitosan/carbon nanotube dispersion. It has been also observed that the homogenous fibers with an average diameter of 275 nm could be prepared with a conductivity of 9×10^{-5}.

KEYWORDS

- **Carbon nanotubes**
- **Chitin**
- **Chitosan**
- **Dynamic light scattering**

Chapter 8

Progress in Production of Nanofiber Web

M. Kanafchian and A. K. Haghi

INTRODUCTION

Clothing is a person's second skin, since it covered great parts of the body and having a large surface area in contact with the environment. Therefore, clothing is proper interface between environment and human body, and could act as an ideal tool to enhance personal protection. Over the years, growing concern regarding health and safety of persons in various sectors, such as industries, hospitals, research institutions, battlefields, and other hazardous conditions has led to intensive research and development in field of personal protective clothing. Nowadays, there are different types of protective clothing. The simplest and most preliminary of this equipment is made from rubber or plastic that is completely impervious to hazardous substances, air and water vapor. Another approach to protective clothing is laminating activated carbon into multilayer fabric in order to absorb toxic vapors from environment and prevent penetration to the skin [1]. The use of activated carbon is considered only a short term solution because it loses its effectiveness upon exposure to sweat and moisture. The use of semi-permeable membranes as a constituent of the protective material is another approach. In this way, reactive chemical decontaminants encapsulates in microparticles [2] or fills in microporous Hollow fibers [3]. The microparticle or fiber walls are permeable to toxic vapors, but impermeable to decontaminants, so that the toxic agents diffuse selectively into them and neutralize. All of these equipments could trap such toxic pollutions but usually are impervious to air and water vapor, and thus retain body heat. In other words, a negative relationship always exists between thermal comfort and protection performance for currently available protective clothing. For example, nonwoven fabrics with high air permeability exhibit low barrier performance, whereas microporous materials, laminated fabrics, and tightly constructed wovens offer higher level of protection but lower air permeability. Thus, there still exists a very real demand for improved protective clothing that can offer acceptable levels of impermeability to highly toxic pollutions of low molecular weight, while minimizing wearer discomfort and heat stress [4].

Electrospinning provides an ultrathin membrane-like web of extremely fine fibers with very small pore size and high porosity, which makes them excellent candidates for use in filtration, membrane, and possibly protective clothing applications. Preliminary investigations have indicated that the using of nanofiber web in protective clothing structure could present minimal impedance to air permeability and extremely efficiency in trapping aerosol toxic pollutions. Potential of electrospun webs for future protective clothing systems has been investigated [5–7]. Schreuder-Gibson et al has shown an enhancement of aerosol protection via a thin layer of electrospun fibers.

They found that the electrospun webs of nylon 66, polybenzimidazole, polyacryloni-trile, and polyurethane provided good aerosol particle protection, without a consider-able change in moisture vapor transport or breathability of the system [5]. While nano-fiber webs suggest exciting characteristics, it has been reported that they have limited mechanical properties [8, 9]. To compensate this drawback in order to use of them in protective clothing applications, electrospun nanofiber webs could be laminated via an adhesive into a multilayer fabric system [10–11]. The protective clothing made of this multilayer fabric will provide both protection against toxic aerosol and thermal comfort for user.

The adhesives in the fabric lamination are as solvent/water-based adhesive or as hot-melt adhesive. At the first group, the adhesives are as solution in solvent or wa-ter, and solidify by evaporating of the carrying liquid. Solvent-based adhesives could "wet" the surfaces to be joined better than water-based adhesives, and also could so-lidify faster. But unfortunately, they are environmentally unfriendly, usually flamma-ble, and more expensive than those. Of course it does not mean that the water-based adhesives are always preferred for laminating, since in practice, drying off water in terms of energy and time is expensive too. Besides, water-based adhesives are not re-sisting to water or moisture because of their hydrophilic nature. At the second group, hot-melt adhesives are environmentally friendly, inexpensive, require less heat and energy, and so are now more preferred. Generally, there are two procedures to melt these adhesives; static hot-melt laminating that accomplish by flat iron or Hoffman press and continuous hot-melt laminating that uses the hot calendars. In addition, these adhesives are available in several forms; as a web, as a continuous film, or in powder form. The adhesives in film or web form are more expensive than the corresponding adhesive powders. The web form are discontinuous and produce laminates which are flexible, porous, and breathable, whereas, continuous film adhesives cause stiffening and produce laminates which are not porous and permeable to both air and water va-por. This behavior attributed to impervious nature of adhesive film and its shrinkage under the action of heat [12]. Thus, the knowledge of laminating skills and adhesive types is very essential to producing an appropriate multilayer fabric. Specifically, this subject becomes more highlight as we will laminate the ultrathin nanofiber web into multilayer fabric, because the laminating process may be adversely influenced on the nanofiber web properties. Lee et al [7], without disclosure of laminating details, re-ported that the hot-melt method is more suitable for nanofiber web laminating. In this method, laminating temperature is one of the most effective parameters. Incorrect selection of this parameter may lead to change or damage ultrathin nanofiber web. Therefore, it is necessary to find out a laminating temperature which has the least ef-fect on nanofiber web during process.

The purpose of this study is to consider the influence of laminating temperature on the nanofiber web/multilayer fabric properties to make protective fabric which is resistance against aerosol pollutions. Multilayer fabrics were made by laminating of nanofiber web into cotton fabric via hot-melt method at different temperatures. Effects of laminating temperature on the nanofiber web morphology, air transport properties, and the adhesive force were discussed.

EXPERIMENTAL

Electrospinning and Laminating Process

The electrospinning conditions and layers properties for laminating are summarized in Table 8.1. Polyacrylonitrile (PAN) of 70,000 g/mol molecular weight from Poly-acryl Co. (Isfehan, Iran) has been used with N, N-dimethylformamide (DMF) from Merck, to form a 12 % Wt polymer solution after stirring for 5 hr and exposing for 24 hr at ambient temperature. The yellow and ripened solution was inserted into a plastic syringe with a stainless steel nozzle and then it was placed in a metering pump from WORLD PRECISION INSTRUMENTS (Florida, USA). Next, this set installed on a plate which it could traverse to left-right direction along drum collector (Fig. 8.1). The electrospinning process was carried out for 8 hr and the nanofibers were collected on an aluminum-covered rotating drum which was previously covered with a Poly-Propylene Spun-bond Nonwoven (PPSN) substrate. After removing of PPSN covered with nano-fiber from drum and attaching another layer of PPSN on it, this set was incorporated between two cotton weft-warp fabrics as a structure of fabric-PPSN-nanofiber web-PPSN-fabric (Fig. 8.2). Finally, hot-melt laminating performed using a simple flat iron for one minute, under a pressure of 9gf/cm^2 and at temperatures 85, 110, 120, 140, and 150°C (above softening point of PPSN) to form the multilayer fabrics.

Table 8.1. Electrospinning conditions and layers properties for laminating.

Electrospinning conditions		Layer properties	
		PPSN	
Polymer concentration	12% w/w	Thickness	0.19 mm
Flow rate	1μl/h	Air permeability	824 cm^2/s/cm^2
		Melting point	140°C
		Mass	25 g/m^2
Nozzle inner diameter	0.4 mm	Mass	25 g/m^2
Nozzle-Drum distance	7 cm	Nanofiber web Mass	3.82 g/m^2
Voltage	11 KV	Fabric	
Drum speed	9 m/min	Thickness	0.24 mm
Spinning Time	8 hr	Warp-weft density	25×25 per cm

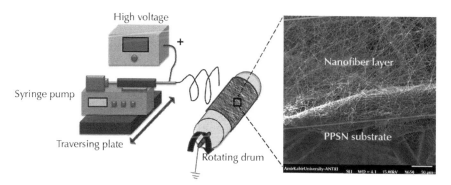

Figure 8.1. Electrospinning setup and an enlarged image of nanofiber layer on PPSN.

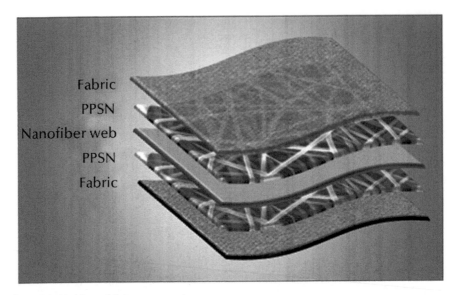

Figure 8.2. Multilayer fabric components.

Nanofiber Web Morphology

A piece of multilayer fabrics were freeze fractured in liquid nitrogen and after sputter-coating with Au/Pd, a cross-section image of them captured using a scanning electron microscope (Seron Technology, AIS-2100, Korea).

Also, to consider the nanofiber web surface after hot-melt laminating, other laminations were prepared by a non-stick sheet made of Teflon (0.25 mm thickness) as a replacement for one of the fabrics (fabric /PPSN/nanofiber web/PPSN/Teflon sheet). Laminating process was carried out at the same conditions which mentioned to produce primary laminations. Finally, after removing of Teflon sheet, the nanofiber layer side was observed under an optical microscope (MICROPHOT-FXA, Nikon, Japan) connected to a digital camera.

Measurement of Air Permeability

Air permeability of multilayer fabrics after laminating were tested by fabric air permeability tester (TEXTEST FX3300, Zürich, Switzerland). It was tested 5 pieces of each sample under air pressure of 125 Pa, at ambient condition (16°C, 70% RH) and then obtained average air permeability.

RESULTS AND DISCUSSION

In electrospinning phase, PPSN was chosen as a substrate to provide strength to the nanofiber web and to prevent of its destruction in removing from the collector. In Fig. 8.1, an ultrathin layer of nanofiber web on PPSN layer is illustrated, which conveniently shows the relative fiber sizes of nanofibers web (approximately 380 nm) compared to PPSN fibers. Also, this figure shows that the macropores of PPSN substrate is covered with numerous electrospun nanofibers, which will create innumerable micro-scopic pores in this system. But in laminating phase, this substrate acts as an adhesive and causes to bond the nanofiber web to the fabric. In general, it is relatively simple to create a strong bond between these layers, which guarantees no delamination or fail-ure in multilayer structures; the challenge is to preserve the original properties of the nanofiber web and fabrics to produce a laminate with the required appearance, handle, thermal comfort, and protection. In other words, the application of adhesive should have minimum affect on the fabric flexibility or on the nanofiber web structure. In order to achieve to this aim, it is necessary that: a) the least amount of a highly effec-tive adhesive applied, b) the adhesive correctly cover the widest possible surface area of layers for better linkage between them, and c) the adhesive penetrate to a certain extent of the nanofiber web/fabric [12]. Therefore, we selected PPSN, which is a hot-melt adhesive in web form. As mentioned above, the perfect use of web form adhesive can be lead to produce multilayer fabrics which are porous, flexible, and permeable to both air and water vapor. On the other hand, since the melting point of PPSN is low, hot-melt laminating can perform at lower temperatures. Hence, the probability of shrinkage that may happen on layers in effect of heat becomes smaller. Of course in this study, we utilized cotton fabrics and polyacrylonitrile nanofiber web for laminat-ing, which intrinsically are resistant to shrinkage even at higher temperatures (above laminating temperature). By this description, laminating process performed at five dif-ferent temperatures to consider the effect of laminating temperature on the nanofiber web/multilayer fabric properties.

Figure 8.3(A–E) shows a SEM image of multilayer fabric cross-section after lami-nating at different temperatures. It is obvious that these images do not deliver any information about nanofiber web morphology in multilayer structure, so it becomes impossible to consider the effect of laminating temperature on nanofiber web. There-fore, in a novel way, we decided to prepare a secondary multilayer by substitution of one of the fabrics (ref. Fig. 8.2) with Teflon sheet. By this replacement, the surface of nanofiber web will become accessible after laminating; because Teflon is a non-stick material and easily separates from adhesive.

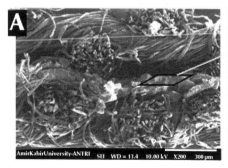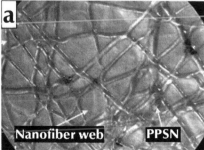

Figure 8.3. *(Continued)*

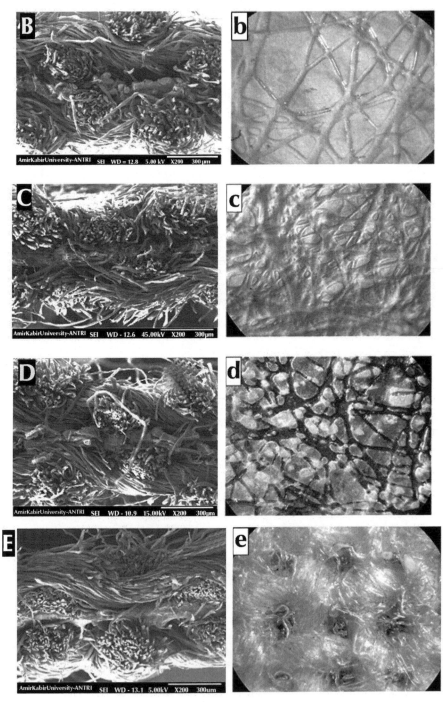

Figure 8.3. SEM images of multilayer fabric cross-section at 200 magnification (A–E) and optical microscope images of nanofiber web surface at 100 magnification (a–e).

Fig. 8.3(a–e) presents optical microscope images of nanofiber web and adhesive after laminating at different temperatures. It is apparent that the adhesive gradually flattened on nanofiber web (Fig. 8.3 (a–c)) when laminating temperature increased to melting point of adhesive (140°C). This behavior is attributed to increment in plasticity of adhesive because of temperature rise and the pressure applied from the iron weight. But, by selection of melting point as laminating temperature, the adhesive completely melted and began to penetrate into the nanofiber web structure instead of spread on it (Fig. 8.3(d)). This penetration, in some regions, was continued to some extent that the adhesive was even passed across the web layer. The dark crisscross lines in Fig. 8.3 (d) obviously show where this excessive penetration is occurred. The adhesive penetration could intensify by increasing of laminating temperature above melting point; because the fluidity of melted adhesive increases by temperature rise. Fig. 8.3 (e) clearly shows the amount of adhesive diffusion in the web which was laminated at 150°C. At this case, the whole diffusion of adhesive lead to create a transparent film and to appear the fabric structure under optical microscope.

Also, to examine how laminating temperature affect the breathability of multilayer fabric, air permeability experiment was performed. The bar chart in Fig. 8.4 indicates the effect of laminating temperature on air transport properties of multilayer fabrics. As might be expected, the air permeability decreased with increasing laminating temperature. This procedure means that the air permeability of multilayer fabric is related to adhesive's form after laminating, because the polyacrylonitrile nanofiber web and cotton fabrics intrinsically are resistant to heat (ref. Fig. 8.3). Of course, it is to be noted that the pressure applied during laminating can leads to compact the web/fabric structure and to reduce the air permeability too. Nevertheless, this parameter did not have effective role on air permeability variations at this work, because the pressure applied for all samples had the same quantity. As discussed, by increasing of laminating temperature to melting point, PPSN was gradually flattened between layers so that it was transformed from web-form to film-like. It is obvious in Fig. 8.3 (a–c) that the pore size of adhesive layer becomes smaller in effect of this transformation. Therefore, we can conclude that the adhesive layer as a barrier resists to convective air flow during experiment and finally reduces the air permeability of multilayer fabric according to the pore size decrease. But, this reason was not acceptable for the samples that were laminated at melting point (140°C); since the adhesive was missed self layer form because of penetration into the web/fabric structures (Fig. 8.3 d). At these samples, the adhesive penetration leads to block the pores of web/fabrics and to prevent of the air pass during experiment. It should be noted that the adhesive was penetrated into the web much more than the fabric, because PPSN structurally had more surface junction with the web (Fig. 8.3 (A–E)). Therefore, at here, the nanofiber web contained the adhesive itself could form an impervious barrier to air flow.

Furthermore, we only observed that the adhesive force between layers was improved according to temperature rise. For example, the samples laminated at 85°C were exhibited very poor adhesion between the nanofiber web and the fabrics as much as they could be delaminated by light abrasion of thumb. Generally, it is essential that no delamination occurs during use of this multilayer fabric, because the nanofiber web might be destroyed due to abrasion of fabric layer. Before melting point, improving the adhesive force according to temperature rise is simultaneously attributed to the more penetration of adhesive into layers and the expansion of bonding area between them, as already discussed. Also at melting point, the deep penetration of adhesive into the web/fabric leads to increase in this force.

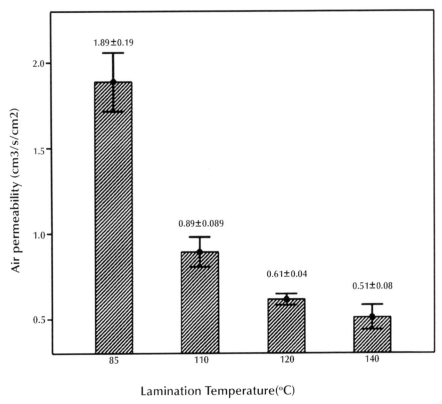

Figure 8.4. Air permeability of multilayer fabric after laminating at different temperatures.

SUMMARY

In this study, the effect of laminating temperature on the nanofiber web/multilayer fabric properties investigated to make next-generation protective clothing. First, we demonstrate that it is impossible to consider the effect of laminating temperature on the nanofiber web morphology by a SEM image of multilayer fabric cross-section. Thus, we prepared a surface image of nanofiber web after laminating at different temperature using an optical microscope. It was observed that nanofiber web was approximately unchanged when laminating temperature was below PPSN melting point. In addition, to compare air transport properties of multilayer fabrics, air permeability tests were performed. It was found that by increasing laminating temperature, air permeability was decreased. Furthermore, it only was observed that the adhesive force between layers in multilayer fabrics was increased with temperature rise. These results indicate that laminating temperature is an effective parameter for laminating of nanofiber web into fabric structure. Thus, varying this parameter could lead to developing fabrics with different levels of thermal comfort and protection depending on our need and use. For example, laminating temperature should be selected close to melting point of adhesive, if we would produce a protective fabric with good adhesive force and medium air permeability.

KEYWORDS

- **Hot-melt adhesives**
- **Nanofiber web**
- **Poly-propylene spun-bond nonwoven**
- **Semi-permeable membranes**
- **Solvent/water-based adhesive**

References

1

1. Huang, Z. M., Zhang, Y. Z., Kotaki, M., and Ramakrishna, S. (2003). A review on polymer nanofibers by electrospinning and their applications in nanocomposites. *Composites Science and Technology* **63**, 2223–2253.

2. Ramakrishna, S., Fujihara, K., Teo, W. E., Lim, T. C., and Ma, Z. (2005). *An Introduction to Electrospinning and Nanofibers*. World scientific, Singapore.

3. Fennessey, S. F. and Farris, R. J. (2004). Fabrication of aligned and molecularly oriented electrospun polyacrylonitrile nanofibers and the mechanical behavior of their twisted yarn. *Polymer* **45**, 4217–4225.

4. Pan, H., Li, L., Hu, L., and Cui, X. (2006). Continuous aligned polymer fibers produced by a modified electrospinning method. *Polymer* **47**, 4901–4904.

5. Zussman, E., Chen, X., Ding, W., Calabri, L., Dikin, D. A., Quintana, J. P., and Ruoff, R.S. (2005). Mechanical and structural characterization of electrospun PAN-derived carbon nanofibers. *Carbon* **43**, 2175–2185.

6. Gu, S. Y., Ren, J., and Wu, Q. L. (2005). Preparation and structures of electrospun PAN nanofibers as a precursor of carbon nanofibers. *Synthetic Metals* **155**, 157–161.

7.. Jalili, R., Morshed, M., and Hosseini Ravandi, S. A. (2006). Fundamental parameters affecting electrospinning of PAN nanofibers as uniaxially aligned fibers. *Journal of Applied Polymer Science* **101**, 4350–4357.

8. Causin, V., Marega, C., Schiavone, S., and Marigo, A. (2006). A quantitative differentiation method for acrylic fibers by infrared spectroscopy. *Forensic Science International* **151**, 125–131.

9. Mathieu, D. and Grand, A. (1988). *Ab initio* Hartree-Fock Raman spectra of Polyacrylonitrile. *Polymer* **39**, 5011–5017.

10. Huang, Y. S. and Koenig, J. L. (1971). Raman spectra of polyacrylonitrile. *Applied Spectroscopy* **25**, 620–622.

11. Liu, T. and Kumar, S. (2003). Quantitative characterization of SWNT orientation by polarized Raman spectroscopy. *Chemical Physics Letters* **378**, 257–262.

12. Jones, W. J., Thomas, D. K., Thomas, D. W., and Williams, G. (2004). On the determination of order parameters for homogeneous and twisted nematic liquid crystals from Raman spectroscopy. *Journal of Molecular Structure* **708**, 145–163.

13. Davidson, J. A., Jung, H. T., Hudson, S. D., and Percec, S. (2000). Investigation of molecular orientation in melt-spun high acrylonitrile fibers. *Polymer* **41**, 3357–3364.

14. Soulis, S. and Simitzis, J. (2005). Thermomechanical behaviour of poly acrylonitrile-co-(methyl acrylate). fibres oxidatively treated at temperatures up to 180°C. *Polymer International* **54**, 1474–1483.

15. Rahaman, M. S. A., Ismail, A. F., and Mustafa, A. (2007). A review of heat treatment on polyacrylonitrile fiber. *Polymer Degradation and Stability* **92**, 1421–1432.

16. Sen, K., Bajaj, P., and Sreekumar, T. V. (2003). Thermal behavior of drawn acrylic fibers. *Journal of Polymer Science: Part B: Polymer Physics* **41**, 2949–2958.

2

1. Setti, L., Fraleoni-Morgera, A., Ballarin, B., Filippini, A., Frascaro, D., and Piana, C. (2005). An amperometric glucose biosensor prototype fabricated by thermal inkjet printing. *Biosensors and Bioelectronics* **20**, 2019–2026.

2. Ren, X., Meng, X., and Tang, F. (2005). Preparation of Ag-Au nanoparticle and its application to glucose biosensor. *Sensors and Actuators B* **110**, 358–363.

3. Foxx, D. and Kalu, E. (2007). Amperometric biosensor based on thermally activated polymer-stabilized metal nanoparticles. *Electrochemistry Communications* **9**, 584–590.

4. Lin, Z., Huang, L., Liu, Y., Lin, J. M., Chi, Y., and Chen, G. (2008). Electrochemiluminescent biosensor based on multi-wall carbon nanotube/nano-Au modified electrode. *Electrochemistry Communications* **10**, 1708–1711.

5. Chu, X., Wu, B., Xiao, C., Zhang, X., and Chen, J. (2010). A new amperometric glucose biosensor based on platinum nanoparticles/polymerized ionic liquid-carbon nanotubes nanocomposites. *Electrochimica Acta* **55**, 2848–2852.

6. Wang, Z., Ai, F., Xu, Q., Yang, Q., Yu, J. H., Huang, W. H., and Zhao, Y. D. (2010). Electrocatalytic activity of salicylic acid on the platinum nanoparticles modified electrode by electrochemical deposition. *Colloids and Surfaces B: Biointerfaces* **76**, 370–374.

7. Qiu, J. D., Peng, H. P., Liang, R. P., and Xia, X. H. (2010). Facile preparation of magnetic core–shell Fe3O4@Au nanoparticle/myoglobin biofilm for direct electrochemistry. *Biosensors and Bioelectronics* **25**, 1447–1453.

8. Ragupathy, D., Gopalan, A. L., and Lee, K. P. (2010). Electrocatalytic oxidation and determination of ascorbic acid in the presence of dopamine at multiwalled carbon nanotube–silica network–gold nanoparticles based nanohybrid modified electrode. *Sensors and Actuators B* **143**, 696–703.

9. Yan, S., He, N., Song, Y., Zhang, Z., Qian, J., and Xiao, Z. (2010). A novel biosensor based on gold nanoparticles modified silicon nanowire arrays. *Journal of Electroanalytical Chemistry* **641**, 136–140.

10. Wang, H., Wang, X., Zhang, X., Qin, X., Zhao, Z., Miao, Z., Huang, N., and Chen, Q. (2009). A novel glucose biosensor based on the immobilization of glucose oxidase onto gold nanoparticles-modified Pb nanowires. *Biosensors and Bioelectronics* **25**, 142–146.

11. Moreno, M., Rincon, E., Pérez, J. M., González, V. M., Domingo, A., and Dominguez, E. (2009). Selective immobilization of oligonucleotide-modified gold nanoparticles by electrodeposition on screen-printed electrodes. *Biosensors and Bioelectronics* **25**, 778–783.

12. Chen, M. and Diao, G. (2009). Electrochemical study of mono-6-thio-β-cyclodextrin/ferrocene capped on gold nanoparticles: Characterization and application to the design of glucose amperometric biosensor. *Talanta* **80**, 815–820.

13. Azzam, E. M. S., Bashir, A., Shekhah, O., Alawady, A. R. E., Birkner, A., Grunwald, C., and Wöll, C. H. (2009). Fabrication of a surface plasmon resonance biosensor based on gold nanoparticles chemisorbed onto a 1, 10-decanedithiol self-assembled monolayer. *Thin Solid Films* **518**, 387–391.

14. Ren, X., Meng, X., Chen, D., Tang, F., and Jiao, J. (2005). Using silver nanoparticle to enhance current response of biosensor. *Biosensors and Bioelectronics* **21**, 433–437.

15. Ma, S., Lu, W., Mu, J., Liu, F., and Jiang, L. (2008). Inhibition and enhancement of glucose oxidase activity in a chitosan-based electrode filled with silver nanoparticles. *Colloids and Surfaces A: Physicochem, Engineering Aspects* **324**, 9–13.

16. Lin, J., He, C., Zhao, Y., and Zhang, S. (2009). One-step synthesis of silver nanoparticles/carbon nanotubes/chitosan film and its application in glucose biosensor. *Sensors and Actuators B* **137**, 768–773.

17. Shim, I., Lee, Y., Lee, K., and Joung, J. (2008). An organometallic route to highly monodispersed silver nanoparticles and their application to ink-jet printing. *Materials Chemistry and Physics* **110**, 316–321.

18. Lee, Y., Choi, J. R., Lee, K. J., Stott, N. E., and Kim, D. (2008). Large-scale synthesis of copper nanoparticles by chemically controlled reduction for applications of inkjet-printing electronics. *Nanotechnology* **19**, 41.

19. Bidoki, S. M., Nouri, J., and Heidari, A. A. (2007). Inkjet deposited circuit components. *Journal of Micromechanics and Microengineering* **17**, 967–974.

20. Xian, Y. (2006). Glucose biosensor based on Au nanoparticles-conductive polyaniline nanocomposite. *Biosensors and Bioelectronics* **21**, 1996–2000.

21. Wang, J. (2008). Disposable biosensor based on immobilization of glucose oxidase at gold nanoparticles electrodeposited on indium tin oxide electrode. *Sensors and Actuators B* **135**, 283–288.

3

1. Guo, R. H., Jiang, S. Q., Yuen, C. W. M., and Ng, M. C. F. (2008). An alternative process for electroless copper plating on polyester fabric. *Journal of Material Science: Mater Electron*, doi 10.1007/s10854-008-9594-4.

2. Lin, Y. M. and Yen, S. H. (2001). Effects of additives and chelating agents on electroless copper plating. *Applied Surface Science* **178**, 116–126.

3. Xueping, G., Yating, W., Lei, L., Bin, Sh., and Wenbin, H. (2008). Electroless plating of Cu–Ni–P alloy on PET fabrics and effect of plating parameters on the properties of conductive fabrics. *Journal of Alloys and Compounds* **455**, 308–313.

4. Li, J., Hayden, H., and Kohl, P. A. (2004). The influence of 2,2'-dipyridyl on non-formaldehyde electroless copper plating. *Electrochim Acta* **49**, 1789–1795.

5. Larhzil, H., Cisse, M., Touir, R., Ebn Touhami, M., and Cherkaoui, M. (2007). Electrochemical and SEM investigations of the influence of gluconate on the electroless deposition of Ni–Cu–P alloys. *Electrochimica Acta* **53**, 622–628.

6. Gaudiello, J. G. and Ballard, G. L. (1993). Mechanistic insights into metal-mediated electroless copper plating Employing hypophosphite as a reducing agent. *IBM Journal of Research and Development* **37**, 107–116.

7. Xueping, G., Yating, W., Lei, L., Bin, Sh., and Wenbin, H. (2007). Electroless copper plating on PET fabrics using hypophosphite as reducing agent. *Surface & Coatings Technology* **201**, 7018–7020.

8. Han, E. G., Kim, E. A., and Oh, K. W. (2001). Electromagnetic interference shielding effectiveness of electroless Cu-platted PET fabrics. *Synthetic Metals* **123**, 469–476.

9. Jiang, S. Q. and Guo, R. H. (2008). Effect of polyester fabric through electroless Ni-P plating. *Fibers and Polymers* **9**, 755–760.

10. Djokic, S. S. (2002). Metal deposition without an external current. In *Fundamental aspects of electrometallurgy*. K. I. Popov, S. S. Djokic, and B. N. Grgur (Eds.). Kluwer Academic Publishers, New York, Chapter 10, pp. 249–270.

11. Mallory, O. and Hajdu, B. J (Eds.) (1990). *Eleolesctrs Plating: Fundamentals and Applications*. Noyes Publication, New York.

12. Agarwala, R. C. and Agarwala, V. (2003). Electroless alloy/composite coatings: A review. *Sadhana* **28**, 475–493.

13. Paunovic, M. (1968). Electrochemical aspects of electroless deposition of metals. *Plating* **51**, 1161–1167.

14. Chang, H. F. and Lin, W. H. (1998). TPR study of electroless plated copper catalysts. *Korean Journal of Chemical Engineering* **15**, 559–562.

4

1. Mäntysalo, M. and Mansikkamäki, P. (2009). An inkjet-deposited antenna for 2.4 GHz applications, AEU. *International Journal of Electronics and Communications* **63**, 31–35.

2. Das, R. and Harrop, P. (2009). Printed, organic & flexible electronics forecasts, players & opportunities 2009–2029. IDTechex Webpage, Retrieved from http://www.IDTechex.com

3. Bidoki, S. M., McGorman, D., Lewis, D. M., Clark, M., Horler, G., and Miles, R. E. (2005). Inkjet printing of conductive patterns on textile fabrics. *The AATCC Review* **5**, 11–13.

4. Derby, B. and Reis, N. (2003). Inkjet printing of highly loaded particulate suspensions, Materials Research Society, Vol. 28.

5. Radivojevic, Z., Andersson, K., Hashizume, K., Heino, M., Mantysalo, M., Mansikkamaki, P., Matsuba, Y., and Terada, N. (2006). Optimised curing of silver ink jet based printed traces, TIMA Editions/ THERMINIC 2006, Nice, Côte d'Azur, France, September 27–29, 2006.

6. Allen, M. L., Aronniemi, M., Mattila, T., Alastalo, A., Ojanper, K., Suhonen, M., and Seppa, H. (2008). Electrical sintering of nanoparticle structures. *Nanotechnology* **19**.

7. Johnson, R. D. and Damarel, W. N. (2003). Depositing solid materials, World Patent WO03049515.

8. Bidoki, S. M., Lewis, D. M., Clark, M., Vakurov, V. A., Millner, P. A., and McGorman, D. (2007). Inkjet fabrication of

electronic components. *Journal of Micromechanics and Microengineering* **17**, 967–974.

9. Hochberg, J. and Foster, P. (2006). Four point probe I–V electrical measurements using the Zyvex test system employing a Keithley 4200, Zyvex Corporation, Retrieved from http://www.zyvex.com/Documents/9702.PDF

5

1. Ren, Y., Zhang, H., Huang, H., Wang, X., Zhou, Z., Cui, G., and An, Y. (2009). *In vitro* behavior of neural stem cells in response to different chemical functional groups. *Biomaterial* **30**, 1036–1044.

2. Lindsay, R. M. (1988). Nerve growth factors (NGF, BDNF) enhance axonal regeneration but are not required for survival of adult sensory neurons. *Journal of Neuroscience* **8**, 3337–3342.

3. Kimpinski, K., Campenot, R. B., and Mearow, K. (1997). Effects of the neurotrophins nerve growth factor, neurotrophin-3, and brain-derived neurotrophic factor (BDNF) on neurite growth from adult sensory neurons in compartmented cultures. *Journal of Neurobiology* **33**, 395–410.

4. Noble, J., Munro, C. A., Prasad, V. A., and Midha, R. (1988). Analysis of upper and lower extremity peripheral nerve injuries in a population of patients with multiple injuries. *Journal of Trauma* **45**, 116–122.

5. Archibald, S. J., Krarup, C., Shefner, J., Li, S. T., and Madison, R. D. (1988). A collagen-based nerve guide conduit for peripheral nerve repair: An electrophysiological study of nerve regeneration in rodents and nonhuman primates. *Journal of Comparative Neurology* **306**, 685–696.

6. Willerth, S. and Sakiyama-Elbert, S. (2007). Approaches to neural tissue engineering using scaffolds for drug delivery. *Advanced Drug Delivery Review* **59**, 325–338.

7. Uebersax, L., Mattotti, M., Papaloizos, M., Merkle, H., Gander, B., and Meinel, L. (2007). Silk fobroin matrices for the controlled release of nerve growth factor (NGF). *Biomaterials* **28**, 4449–4460.

8. Weber, R. A., Breidenbach, W. C., Brown, R. E., Jabaley, M. E., and Mass, D. P. (2000). A randomized prospective study of polyglycolic acid conduits for digital nerve reconstruction in humans. *Plastic Reconstruction Surgery* **106**, 1036–1045.

9. Wang, X., Hu, W., Cao, Y., Yao, J., Wu, J., and Gu, X. (2005). Dog sciatic nerve regeneration across a 30-mm defect bridged by a chitosan/PGA artificial nerve graft. *Brain* **128**, 1897–1910.

10. Engel, E., Michiardi, A., Navarro, M., and Laroix, D. (2007). Nanotechnology in regenerative medicine: the materials side. *Trends Biotechnology* **26**, 39–47.

11. Agarwal, S., Wendroff, J., and Greiner, A. (2008). Use of electrospinning technique for biomedical applications. *Polymer* **49**, 5603–5621.

12. Gandhi, M., Yang, H., Shor, L., and Ko, F. (2009). Post-spinning modification of electrospun nanofiber nanocomposites from *Bombyx mori* silk and carbon nanotubes. *Polymer* **50**, 1918–1924.

13. Ma, P. X. (2008). Biomimetic materials for tissue engineering. *Advanced Drug Delivery Review* **60**, 184–198.

14. Khademhosseini, A., Langer, R., Borenstein, J., and Vacanti, J. (2006). Microscale technologies for tissue engineering and biology. *Proceeding of National Academy of Science* **21**, 2480–2487.

15. Travis, J., Horst, A., and Recum, V. (2008). Electrospinning: Application in drug delivery and tissue engineering. *Biomater* **29**, 1989–2006.

16. Chew, S., Mi, R., Hoke, A., and Leong, K. (2007). Aligned protein–polymer composite fiber enhance nerve regeneration: A potential tissue–engineering platform. *Advanced Functional Materials* **17**, 1288–1269.

17. Drury, J. and Moony, D. (2003). Hydrogels for tissue engineering: scaffold design variables and applications. *Biomaterials* **24**, 4337–4351.

18. Costa-Junior, E., Barbosa-Stacioli, E., Mansur, A., Vasconcelos, W., and Mansur, H. (2008). Preparation and characterization of chitosan/poly(vinyl alcohol) chemically crosslinked blends for biomedical application. *Carbohydrate Polymer* **77**, 718–724.

19. Chuachamsai, A., Lertviriyasawat, S., and Danwanichakul, P. (2008). Spinnability and defect formation of chitosan/poly vinyl alcohol electrospun nanofibers. *Thammasat Inernational Journal of Science and Technology* **13**, 24–29.

20. Duan, B., Wu, L., Li, X., Yuan, X., and Yao, K. (2007). Degredation of electrospun PLGA-chitosan/PVA membranes and their cytocompatibility *in vitro*. *Biomaterial Science* **18**, 95–115.

21. Kumar, A., Depan, D., Tomer, N., and Singh, R. (2009). Nanoscale particles for polymer degradation and stabilization—Trends and future perspectives. *Progress in Polymer Science* **34**, 479–515.

22. Lee, K., Jeong, L., Kang, Y., Lee, S., and Park, W. (2009). Electrospinning of polysaccharides for regenerative medicine. *Advanced Drug Delivery Review* **61**, 1020–1032.

23. Mansur, H., Costa, J., Mansur, A., and Stanciolik, B. (2009). Cytocpmpatibility evaluation in cell-culture system of chemically crosslinked chitosan/PVA hydrogels. *Material Science and Engineering C* **12**, 132–140.

24. Mottaghitalab, F., Farokhi, M., Ziabari, M., Divsalar, A., Eslamifar, A., Mottaghitalab, V., and Shokrgozar, M. A. (2010). *Enhancement of neural cell lines proliferation using nano-structured chitosan/poly(vinyl alcohol) scaffolds conjugated with nerve growth factor.* Submitted manuscript to Carbohydrate Polymers.

25. Hollister, J. (2005). Porous scaffold design for tissue engineering. *Natural Materials* **4**, 518–524.

26. Reneker, D. H. and Chun, I. (1996). Conducting polymer fibers of polyaniline doped with camphorsulfonic acid. *Nanotechnology* **7**, 216–230.

27. Harrison, B. S. and Atala, A. (2007). Carbon nanotube applications for tissue engineering. *Biomater* **28**, 344–353.

28. Aliev, A. E. (2008). Bolometric detector on the basis of single-walled carbon nanotube/polymer composite. *Infrared Physics Technology* **16**, 130–138.

29. Abarrategi, A., Gutierrez, M., Moreno-Vicnte, C., Hortiguela, M., Ramos, V., Lopez-Lacomba, J., Ferrer, M., and Monte, F. (2008). Multiwall carbon nanotube scaffolds for tissue engineering purposes. *Biomaterial* **29**, 94–102.

30. Ramires, P. A., Romito, A., Cosentino, F., and Milella, E. (2001). The influence of titinia/hydroxyapatie composite coating on the *in vitro* osteoblast behavior. *Biomaterials* **22**, 1467–1474.

31. Ziabari, M., Mottaghitalab, V., and Haghi, A. K. (2008). Evaluation of electrospun nanofiber pore structure parameters. *Korean Journal of Chemical Engineering* **25**, 923–932.

32. Ziabari, M., Mottaghitalab, V., McGovern, S. T., and Haghi, A. K. (2007). A new image analysis based method for measuring electrospun nanofiber diameter. *Nanoscale Research Letter* **2**, 597–600.

33. Maynard, A. D., Baron, P. A., Foley, M., Shvedova, A. A., Kisin, E. R., and Castranova, V. (2004). Exposure to carbon nanotube materials: Aerosol release during the handling of unrefined single walled carbon nanotube materials. *Toxicology Environment and Health A* **67**, 87–107.

34. Jia, G., Wang, H., Yan, L., Wang, X., Pei, R., Yan, T., et al. (2005). Cytotoxicity of carbon Nanomaterials: Single-wall nanotube, multi-wall nanotube, and fullerene. *Environomental Science and Technology* **39**, 1378–1383.

35. Ilbasmis-Tamer, S., Yilmaz, S., Banoglu, E., and Tuncer Degim, I. (2010). Carbon nanotubes to deliver drug molecules. *Journal of Biomedical Nanotechnology* **6**, 20–27.

36. Senanayake, V., Juurlink, B. H., Zhang, C., Zhan, E., Wilson, L. D., Kwon, J., Yang, J., Lim, Z. L., Brunet, S. M. K., Schatte, G., Maley, J., Hoffmeyer, M. R. E., and Sammynaiken, R. (2008). Do surface defects and modification determine the observed toxicity of carbon nanotubes. *Journal of Biomedical Nanotechnology* **4**, 515–523.

37. Wise, A. J., Smith, J. R., Bouropoulos, N., Yannopoulos, S. N., van der Merwe, S. M., and Fatouros, D. G. (2008). Single-wall carbon nanotube dispersions stabilised with *N*-trimethyl-chitosan. *Journal of Biomedical Nanotechnology* **4**, 67–72.

38. Ruckmani, K., Sankar, V., and Sivakumar, M. (2010). Tissue distribution, pharmacokinetics and stability studies of zidovudine delivered by niosomes and proniosomes. *Journal of Biomedical Nanotechnology* **6**, 43–51.

39. Kato, S., Aoshima, H., Saitoh, Y., and Miwa, N. (2010). Defensive effects of fullerene-C60 dissolved in squalane against the 2,4-nonadienal-induced cell injury in human skin keratinocytes hacat and wrinkle formation in 3D-human skin tissue model. *Journal of Biomedical Nanotechnology* **6**, 52–58.

40. Vallet-Regí, M., Colilla, M., and Izquierdo-Barba, I. (2008). Bioactive mesoporous silicas as controlled delivery systems: Application in bone tissue regeneration. *Journal of Biomedical Nanotechnology* **4**, 1–15.

41. Tätte, T., Kolesnikova, A. L., Hussainov, M., Talviste, R., Lõhmus, R., Romanov, A. E., Hussainova, I., Part, M., and Lõhmus, A. (2010). Crack formation during post-treatment of nano- and microfibres prepared by sol-gel technique. *Journal of Nanoscience and Nanotechnology* **10**, 6009–6016.

42. Chen, J., Li, S., and Chiang, Y. (2010). Bioactive collagen-grafted poly-l-lactic acid nanofibrous membrane for cartilage tissue engineering. *Journal of Nanoscience and Nanotechnology* **10**, 5393–5398.

43. Kim, J. W. and Lee, D. G. (2010). Effect of fiber orientation and fiber contents on the tensile strength in fiber-reinforced composites. *Journal of Nanoscience and Nanotechnology* **10**, 3650–3653.

44. Bispo, V. M., Mansur, A. A. P., Barbosa-Stancioli, E. F., and Mansur, H. S. (2010). Biocompatibility of nanostructured chitosan/poly(vinyl alcohol) blends chemically crosslinked with genipin for biomedical applications. *Journal of Biomedical Nanotechnology* **6**, 166–175.

45. Jones, S. A., Mesgarpour, S., Chana, J., and Forbes, B. (2008). Preparation and characterisation of polymeric nanoparticles using low molecular weight poly(vinyl alcohol). *Journal of Biomedical Nanotechnology* **4**, 319–325.

46. Park, S., Chang, Y., Kim, Y., and Rhee, K. (2010). Anodization of carbon fibers on interfacial mechanical properties of epoxy matrix composites. *Journal of Nanoscience and Nanotechnology* **10**, 117–121.

47. Peng, Y. and Chen, Q. (2009). Ultrasonic-assisted fabrication of highly dispersed copper-multi-walled carbon nanotubes nanowires. *Colloid and Surface A* **12**, 148–156.

48. Pegel, S., Villmow, T., Stoyan, D., and Heinrich, G. (2009). Spatial statistics of carbon nanotube polymer composites. *Polymer* **50**, 2123–2132.

49. Cherukuri, P., Bachilo, S. M., Litovsky, S. H., and Weisman, R. B (2004). Near-infrared fluorescence microscopy of single-walled carbon nanotubes in phagocytic cells. *Journal of American Chemical Society* **126** 15638–15639.

50. Mornet, S., Vasseur, S., Grasset, F., and Duguet, E. (2004). Magnetic nanoparticles design for medical diagnosis and therapy. *Material Chemistry* **14**, 2161–2175.

51. Shaffer, M. and Windle, A. H. (1999). Fabrication and characterization of carbon nanotube/poly(vinyl alcohol) composites. *Advanced Materials* **11**, 937–941.

52. Gong, K., Yan, Y., Zhang, M., Su, L., Xiong, S., and Mao, L. (2005). Electrochemistry and electroanalytical applications of carbon nanotubes: A review. *Analytical Science* **21**, 1383–1391.

6

1. Zywietz, C. (2002). *A brief history of electrocardiography progress through technology.* Biosigna Institute for Biosignal and Systems Research, Hanover, Germany.

2. Abildskov, J. (1989). Electrocardiographic wave form and cardiac arrhythmias. *American Journal of Cardiology* **64**, 29C–31C.

3. Mark, R. G. (2004). Quantitative physiology: Organ transport systems, *Lecture notes from HST/MIT Open Courseware*, Retrieved from http://ocw.mit.edu/OcwWeb/Health-Sciences-and-Technology/HST-542JSpring-2004/Course

4. Gima, K. and Rudy, Y. (2002). Ionic current basis of electrocardiographic waveforms: A model study. *Circulation* **90**, 889–896.

5. He, B., Li, G., and Zhang, X. (2002). Noninvasive three-dimensional activation time

imaging of ventricular excitation by means of a heart-excitation Model. *Physics in Medicine and Biology* **47**, 4063–4078.

6. Carmeliet, E. (1993). Mechanisms and control of repolarization. *Europe Heart Journal* **98**(Suppl. H), 3–13.

7. Oosterom, A. V. (1996). Modeling the cardiac electric field. *Medical and Biological Engineering and Computing* **34**(Suppl. 1), 5–8.

8. Bergey, G. E., Squires, R. D., and Sipple, W. C. (1971). Electrocardiogram recording with pasteless electrodes. *IEEE Transactions on Biomedical Engineering* **18**, 206–211.

9. Cochran, R. J. and Rosen, T. (1980). Contact dermatitis caused by ECG electrode paste. *Southern Medical Journal* **73**, 1667–1671.

10. Wan, S. W. and Nguyen, H. T. (1994). 50 Hz interference and noise in ECG recordings: A review. *Australasian Physical Engineering Science Medicine* **17**, 108–115.

11. Marculescu, D., Marculescu, R., Zamora, N. H., Stanley-Marbell, P., Khosla, P. K., Park, S., Jayaraman, S., Jung, S., Lauterbach, C., Weber, W., Kirstein, T., Cottet, D., Grzyb, J., Troster, G., Jones, M., Martin, T., and Nakad, Z. (2003). Electronic textiles: A platform for pervasive computing. *Proceedings of the IEEE* **91**, 1995–2018.

12. De Rossi, D., Carpi, F., Lorussi, F., Mazzoldi, A., Paradiso, R., Scilingo, E. P., and Tognetti, A. (2003). Electroactive fabrics and wearable biomonitoring devices. *AUTEX Research Journal* **3**, 180–185.

13. Raskovic, D., Martin, T., and Jovanov, E. (2004). Medical monitoring applications for wearable computing. *The Computer Journal* **47**, 495–504.

14. Martin, T., Jovanov, E., and Raskovic, D. (October 2000). Issues in wearable computing for medical monitoring applications: A case study of a wearable ECG monitoring device. *Proceedings of the 4th International Symposium Wearable Computers*, Atlanta, GA, pp. 43–49.

15. Healey, J., Seger, J., and Picard, R. (2001). Quantifying driver stress: Developing a system for collecting and processing biometric signals in natural situations. *Biomedical Science Instruments* **35**, 193–198.

16. Ottenbacher, J., Romer, S., Kunze, C., Grossmann, U., and Stork, W. (2004). Integration of a bluetooth based ECG system into clothing. *Proceedings of the Eighth IEEE International Symposium on Wearable Computers (ISWC'04)*, Arlington, VA, October 31–November 3, pp. 186–187.

17. Bansal, D., Khan, M., and Salhan, A. (2009). A computer based wireless system for online acquisition, monitoring and digital processing of ECG waveforms. *Computers in Biology and Medicine* **39**, 361–367.

18. Coyle, M. (2002). Ambulatory cardiopulmonary data capture. *Proceedings of the Second Annual International IEEE-EMBS Special Topic Conference on Microtechnologies in Medicine and Biology.* Madison, WI, USA, pp. 297–300.

19. MyHeart (2005). Retrieved from http://www.hitech-projects.com/euprojects/myheart/

20. Sensatex (2005). Retrieved from http://www.sensatex.com

21. Verhaert(2005). Retrieved from http://www.verhaert.com

22. Halin, N., Junnila, M., Loula, P., and Aarnio, P. (2005). The LifeShirt system for wireless patient monitoring in the operating room. *Journal of Telemedicine and Telecare* **11**, 41–43.

23. Di Rienzo, M., Rizzo, F., Parati, G., Ferratini, M., Brambilla, G., and Castiglioni, P. (2005). A textile-based wearable system for vital sign monitoring: Applicability in cardiac patients. *Computers in Cardiology* **32**, 699–701.

24. Patel, S. A. and Sciurba, F. C. (2005). Emerging concepts in outcome assessment for COPD clinical trials. *Seminars in Respiratory and Critical Care Medicine* **26**, 253–262.

25. Hande, A., Polk, T., Walker, W., and Bhatia, D. (2006). Self-powered wireless sensor networks for remote patient monitoring in hospitals. *Sensors* **6**, 1102–1117.

26. Milenkovic, A., Otto, C., and Jovanov, E. (2006). Wireless sensor network for personal health monitoring: Issues and an implementation. *Computer Communications* **29**, 2521–2533.

27. Catrysse, M., Puers, R., Hertleer, C., Van Langenhove, L., Van Egmond, H., and Matthys, D. (2004). Towards the integration of textile sensors in a wireless monitoring suit. *Sensors and Actuators A* **114**, 302–311.

28. Huang, C. T., Tang, C. F., Lee, M. C., and Chang, S. H. (2008). Parametric design of yarn-based piezoresistive sensors for smart textiles. *Sensors and Actuators A: Physical* **148**, 10–15.

7

1. Agarwal, S., Wendorff, J. H., and Greiner, A. (2008). *Polymer* **49**, 5603.

2. Li, M., Mondrinos, M. J., Gandhi, M. R., Ko, F. K., Weiss, A. S., and Lelkes, P. I. (2005). *Biomaterials* **26**, 5999.

3. Zeng, J., Yang, L., Liang, Q, Zhang, X., Guan, H., Xu, X., Chen, X., and Jing, X. (2005). *Journal of Controlled Release* **105**, 43.

4. Khil, M.-S., Cha, D.-I., Kim, H.-Y., Kim, I.-S., and Bhattarai, N. (2003). *Journal of Biomedical Materials Research B.* **67B**, 675.

5. Taylor, G. I. (1969). *Proceedings of the Royal Society London* **313**, 453.

6. Doshi, J. and Reneker, D. H. (1995). *Journal of Electrostatics* **35**, 151.

7. Li, D. and Xia, Y. (2004). *Advanced Materials* **16**, 1151.

8. Ziabari, M., Mottaghitalab, V., and Haghi, A. K. (2008). *Korean Journal of Chemical Engineering* **25**, 923.

9. Tan, S. H., Inai, R., Kotaki, M., and Ramakrishna, S. (2005). *Polymer* **46**, 6128.

10. Sukigara, S., Gandhi, M., Ayutsede, J., Micklus, M., and Ko, F. (2003). *Polymer* **44**, 5721.

11. Matthews, J. A., Wnek, G. E., Simpson, D. G., and Bowlin, G. L. (2002). *Biomacromolecules* **3**, 232.

12. McManus, M. C., Boland, E. D., Simpson, D. G., Barnes, C. P., and Bowlin, G. L. (2007). *Journal of Biomedical Materials Research A.* **81A**, 299.

13. Huang, Z.-M., Zhang, Y. Z., Ramakrishna, S., and Lim, C. T. (2004). *Polymer* **45**, 5361.

14. Zhang, X., Reagan, M. R., and Kaplan, D. L. (2009). *Advanced Drug Delivery Reviews* **61**, 988.

15. Noh, H. K., Lee, S. W., Kim, J.-M., Oh, J.-E., Kim, K.-H., Chung, C.-P., Cho, S.-C., Park, W. H., and Min, B.-M. (2006). *Biomaterials* **27**, 3934.

16. Ohkawa, K., Minato, K.-I., Kumagai, G., Hayashi, S., and Yamamoto, H. (2006). *Biomacromolecules* **7**, 3291.

17. Agboh, O. C. and Qin, Y. (1997). *Polymers of Advanced Technology* **8**, 355.

18. Rinaudo, M. (2006). *Progress in Polymer Sciences* **31**, 603.

19. Aranaz, I., Mengíbar, M., Harris, R., Paños, I., Miralles, B., Acosta, N., Galed, G., and Heras, Á. (2009). *Current Chemical Biology* **3**, 203.

20. Neamnark, A., Rujiravanit, R., and Supaphol, P. (2006). *Carbohydrate Polymers* **66**, 298.

21. Duan, B., Dong, C., Yuan, X., and Yao, K. (2004). *Journal of Biomaterials Sciences, Polymer Edition* **15**, 797.

22. Jia, Y.-T., Gong, J. Gu, X.-H., Kim, H.-Y., Dong, J., and Shen, X.-Y. (2007). *Carbohydrate Polymers* **67**, 403.

23. Homayoni, H., Ravandi, S. A. H., and Valizadeh, M. (2009). *Carbohydrate Polymers* **77**, 656.

24. Geng, X., Kwon, O.-H., and Jang, J. (2005). *Biomaterials* **26**, 5427.

25. Torres-Giner, S., Ocio, M. J., and Lagaron, J. M. (2008). *Anglais* **8**, 303.

26. Vrieze, S. D., Westbroek, P., Camp, T. V., and Langenhove, L. V. (2007). *Journal of Materials Science* **42**, 8029.

27. Ohkawa, K., Cha, D., Kim, H., Nishida, A., and Yamamoto, H. (2004). *Macromolecular Rapid Communications* **25**, 1600.

28. Iijima, S. (1991). *Nature* **354**, 56.

29. Esawi, A. M. K. and Farag, M. M. (2007). *Materials & Design* **28**, 2394.

30. Feng, W., Wu, Z., Li, Y., Feng, Y., and Yuan, X. (2008). *Nanotechnology* **19**, 105707.

31. Liao, H., Qi, R., Shen, M., Cao, X., Guo, R., Zhang, Y., and Shi, X. (2011). *Colloids and Surface B*, doi:10.1016/j.colsurfb.2011.02.010.

32. Baek, S.-H., Kim, B., and Suh, K.-D. (2008). *Colloids and Surface A* **316**, 292.

33. Liu, Y.-L., Chen, W.-H., and Chang, Y.-H. (2009). *Carbohydrate Polymers* **76**, 232.

34. Tkac, J., Whittaker, J. W., and Ruzgas, T. (2007). *Biosensors and Bioelectronics* **22**, 1820.

35. Spinks, G. M., Geoffrey, M., Shin, S. R., Wallace, G. G., Whitten, P. G., Kim, S. I., and Kim, S. J. (2006). *Sensors and Actuators B: Chemical* **115**, 678.

36. Zhang, H., Wang, Z., Zhang, Z., Wu, J., Zhang, J., and He, J. (2007). *Advanced Materials* **19**, 698.

37. Deitzel, J. M., Kleinmeyer, J., Harris, D., and Beck Tan, N. C. (2001). *Polymer* **42**, 261.

38. Zhang, S., Shim, W. S., and Kim, J. (2009). *Materials & Design* **30**, 3659.

39. Li, Y., Huang, Z., and Lu, Y. (2006). *European Polymer Journal* **42**, 1696.

8

1. Ziabari, M., Mottaghitalab, V., and Haghi, A. K. (2008). *Korean Journal of Chemical Engineering* **25**, 923.

2. Haghi, A. K. and Akbari, M. (2007). *Physica Status Solidi A* **204**, 1830.

3. Kanafchian, M., Valizadeh, M., and Haghi, A. K. (2011). *Korean Journal of Chemical Engineering* **28**, 428.

4. Ziabari, M., Mottaghitalab, V., and Haghi, A. K. (2008). *Korean Journal of Chemical Engineering* **25**, 905.

5. Kanafchian, M., Valizadeh, M., and Haghi, A. K. (2011). *Korean Journal of Chemical Engineering* **28**, 445.

6. Lee, S. and Obendorf, S. K. (2006). *Journal Applied Polymer Science* **102**, 3430.

7. Lee, S., Kimura, D., Lee, K. H., Park, J. C., and Kim, I. S. (2010). *Textile Research Journal* **80**, 99.

8. Pedicini, A. and Farris, R. J. (2003). *Polymer* **44**, 6857.

9. Lee, K. H., Lee, B. S., Kim, C. H., Kim, H. Y., Kim, K. W., and Nah, C. W. (2005). *Macromolecular Research* **13**, 441.

10. Lee, S. M., Kimura, D., Yokoyama, A., Lee, K. H., Park, J. C., and Kim, I. S. (2009). *Textile Research Journal* **79**, 1085.

11. Liu, L., Huang, Z. M., He, C. L., and Han, X. J. (2006). *Material Science Engineering A* **309**, 435–436.

12. Fung, W. (2002). Materials and their properties. In *Coated and Laminated Textiles*, 1st edition. Woodhead Publishing, pp. 63–71.

Index

Printed and bound by CPI Group (UK) Ltd, Croydon, CR0 4YY

27/10/2024

01779946-0002